RINGS WITH MINIMUM CONDITION

BY

EMIL ARTIN

CECIL J. NESBITT

ROBERT M. THRALL

UNIVERSITY OF MICHIGAN PRESS

ANN ARBOR

Second Printing, 1946
Third Printing, 1948
Fourth Printing, 1952
Fifth Printing, 1955

Paperback ISBN: 978-0-472-75009-2

TABLE OF CONTENTS . . .

CHAPTER I

RINGS AND VECTOR SPACES

1. PRELIMINARY CONCEPTS

We assume at the outset familiarity with the concepts of group, ring, field, and their most elementary properties.* In accordance with currently prevalent usage we mean by "field" always "commutative field." For a (possibly) noncommutative field we shall use the term "sfield."

2. COMPLEX CALCULUS

If $\mathfrak{a}$, $\mathfrak{b}$ are any subsets of an additive group G, we denote by $\mathfrak{c} = \mathfrak{a} + \mathfrak{b}$ the subset of G consisting of the elements $\gamma = \alpha + \beta$ for all α in $\mathfrak{a}$ and β in $\mathfrak{b}$. The set $\mathfrak{c}$ is called the <u>sum</u> of $\mathfrak{a}$ and $\mathfrak{b}$.

More generally, if $\mathfrak{a}_1$, $\mathfrak{a}_2$, ... is any collection of subsets of G, each containing zero, we shall understand by $\mathfrak{a}_1 + \mathfrak{a}_2 + \cdots$ the set of all sums $\alpha_1 + \alpha_2 + \cdots$, where $\alpha_i \varepsilon \mathfrak{a}$ and all but a finite number of the α's are zero. A sum $\Sigma \mathfrak{a}_i$ of subgroups is called <u>direct</u> if the only representation of 0 is $0 + 0 + \cdots$. Observe that the uniqueness of the representation of zero in a direct sum carries with it uniqueness of representation for all elements of the sum.

If $\mathfrak{a}$, $\mathfrak{b}$ are subsets of a ring R, we have in addition to the sum just defined also a <u>product</u> $\mathfrak{a}\mathfrak{b}$, by which we mean the set of all sums $\Sigma\alpha\beta$, where $\alpha \varepsilon \mathfrak{a}$, $\beta \varepsilon \mathfrak{b}$. For repeated multiplication of subsets we have the associative law

$$\mathfrak{a}(\mathfrak{b}\mathfrak{c}) = (\mathfrak{a}\mathfrak{b})\mathfrak{c} = \Sigma\alpha\beta\gamma$$

---

*See, for instance, B. L. van der Waerden, <u>Modern Algebra</u>, Vols. I and II, J. Springer, 1937 and 1940: A. A. Albert, <u>Modern Higher Algebra</u>, The University of Chicago Press, 1937; C. C. MacDuffee, <u>An Introduction to Higher Algebra</u>, John Wiley and Sons, 1940; G. Birkhoff and S. MacLane, <u>A Survey of Modern Algebra</u>, Macmillan, 1941.

For subsets in general we have only the modified distributive laws

$$\mathfrak{a}(\mathfrak{b} + \mathfrak{c}) \subseteq \mathfrak{a}\mathfrak{b} + \mathfrak{a}\mathfrak{c} \quad \text{and} \quad (\mathfrak{a} + \mathfrak{b})\mathfrak{c} \subseteq \mathfrak{a}\mathfrak{c} + \mathfrak{b}\mathfrak{c}.$$

If, however, the subsets all contain 0, the inclusion signs can be replaced by equalities throughout.

A subset $\mathfrak{l}$ of a ring R is said to be a <u>left ideal</u> if (1) $\mathfrak{l}$ is a subgroup (relative to the addition in R) and (2) $R\mathfrak{l} \subseteq \mathfrak{l}$. For a <u>right ideal</u> $\mathfrak{r}$, (2) is replaced by (2') $\mathfrak{r}R \subseteq \mathfrak{r}$. For a <u>two-sided ideal</u> $\mathfrak{o}$, (1), (2), and (2') must all hold. The following theorem is an immediate consequence of the above definitions.

<u>Theorem 1.2A</u>. <u>The sum of any set of ideals (left, right, or two-sided) is again an ideal (left, right, or two-sided)</u>.

3. VECTOR SPACES

A <u>vector space</u> V over a ring R is an additive group with the ring R as operator domain. We write the elements of R as left multipliers of the elements (that is, the <u>vectors</u>) of V. For combination of elements of R with those of V we require that

$$\begin{aligned} \alpha(X + Y) &= \alpha X + \alpha Y \\ (\alpha + \beta)X &= \alpha X + \beta X \qquad (3.1) \\ \alpha(\beta X) &= (\alpha\beta)X . \end{aligned}$$

Let V be a vector space over the sfield $\mathfrak{k}$. The vectors $X_1, \cdots, X_r$ are said to be <u>dependent</u> if there exist elements of $\mathfrak{k}$, $a_1, \cdots, a_r$, not all 0, such that $a_1X_1 + \cdots + a_rX_r = 0$. If no such elements a_i exist, the vectors are called independent. If V contains n independent vectors and every n + 1 vectors in V are dependent, we say that V has $\mathfrak{k}$-dimension n. If no such n exists, V is said to have infinite $\mathfrak{k}$-dimension. Let X be any vector in a $\mathfrak{k}$-space of dimension n, and suppose that $X_1, \cdots, X_n$ are n independent vectors of V. Then we have elements $a, a_1, \cdots, a_n$ in $\mathfrak{k}$ (and not all zero) such that

$$aX + a_1X_1 + \cdots + a_nX_n = 0.$$

The independence of $X_1, \cdots, X_n$ ensures that $a \neq 0$, and so we can solve for X, viz.:

$$X = -a^{-1}a_1X_1 - \cdots - a^{-1}a_nX_n.$$

This shows that any vector in V is a $\mathfrak{k}$-combination of $x_1, \cdots, x_n$.

The existence of $\mathfrak{k}$-dimension for a finitely generated vector space V is a corollary to the following theorem.

Theorem 1.3A. A system of m homogeneous $\mathfrak{k}$-equations $\Sigma_\nu a_{i\nu} x_\nu = 0$ in n unknowns always has a nontrivial solution if $n > m$.

Proof by induction on m. If $m = 0$, there is at least one unknown, and any nonzero value for it satisfies the theorem. Suppose the theorem true for $m - 1$. Now, if $a_{i1} = 0$, $i = 1, \cdots, m$, then $(x_1, \cdots, x_n) = (1, 0, \cdots, 0)$ is a solution of the equations. Otherwise we may arrange the equations so that $a_{11} \neq 0$. Now subtract $a_{i1}a_{11}^{-1}$ times the first equation from the i-th equation, $i = 2, \cdots, m$, giving $m - 1$ equations

$$\Sigma_{\nu=2}^{n} a'_{i\nu} x_\nu = 0 \qquad i = 2, \cdots, m$$

in $x_2, \cdots, x_n$, that is, in $n - 1$ unknowns. Note that $a'_{ik} = a_{ik} - a_{i1}a_{11}^{-1}a_{1k}$. By our induction hypothesis these equations have a nontrivial solution, say $(x_2, \cdots, x_n) = (a_2, \cdots, a_n)$. Now take $a_1 = -a_{11}^{-1}(a_{12}a_2 + \cdots + a_{1n}a_n)$, so that $(a_1, \cdots, a_n)$ satisfies the first equation. For $i > 1$ we have

$$a_{i1}a_1 + a_{i2}a_2 + \cdots + a_{in}a_n = -a_{i1}a_{11}^{-1}(a_{12}a_2$$
$$+ \cdots + a_{1n}a_n) + (a'_{i2} + a_{i1}a_{11}^{-1}a_{12})a_2 + \cdots$$
$$+ (a'_{in} + a_{i1}a_{11}^{-1}a_{1n})a_n = a'_{i2}a_2 + \cdots + a'_{in}a_n = 0,$$

thus completing our induction.

Let $\mathfrak{k}$ be a subfield of a larger field $\mathfrak{k}^*$. Then we can regard $\mathfrak{k}^*$ as a $\mathfrak{k}$-vector space. In this case the $\mathfrak{k}$-dimension of $\mathfrak{k}^*$ is called the degree of $\mathfrak{k}^*$ over $\mathfrak{k}$.

4. COMPLETELY REDUCIBLE VECTOR SPACES

Let V be a vector space with a ring R as left operator domain. We call a subgroup V' of V a subspace (or an R-subspace) if $RV' \subseteq V$ (analogous to definition of left ideal in R). We say that a subspace V' is irreducible if it has as subspaces only itself and the 0-space (that is, the space consisting of the 0-vector alone). The following elementary property of irreducible spaces is used repeatedly.

Lemma 1.4A. If V *is an irreducible vector space over a ring* R *and* X *is any vector in* V, *then* RX = V *or* RX = 0.

The proof is almost trivial. We set $V' = RX$, and show that V' is an R-space. It will then follow from the irreducibility of V that its subspace V' can be only V or 0. By definition, V' is the set of all vectors of the form $X' = \Sigma\alpha_\nu X = \alpha X$, where $\Sigma\alpha_\nu = \alpha$. The equations $\alpha X + \beta X = (\alpha + \beta)X$, $\alpha(\beta X) = (\alpha\beta)X$, which hold for all vectors of V, show that V' is an R-subspace of V.

The concepts of independence and dependence of sets of vectors are somewhat more complicated for spaces over an arbitrary ring than for spaces over a sfield. For instance, it does not follow from a relation $\alpha X + \alpha_1X_1 + \cdots + \alpha_nX_n = 0$ with $\alpha \neq 0$ that X can be expressed as an R-combination of $X_1, \cdots, X_n$. Hence we phrase the definition of dependence in the following way: *the vector* X *is said to be dependent on the vectors* $X_1, \cdots, X_n$ if there is a relation of the form $X = \alpha_1X + \cdots + \alpha_nX_n$ with the α_i in R. We say that *a set of vectors* $X_1, \cdots, X_n$ *is dependent* if any one of them can be expressed as an R-combination of the others.

We say that the nonzero vectors $X_1, \cdots, X_n$ are *independent* if $\alpha_1X_1 + \cdots + \alpha_nX_n = 0$ (with $\alpha_i \in R$) requires $\alpha_1X_1 = \cdots = \alpha_nX_n = 0$. Observe that, according to these definitions, it may happen that a set of vectors is neither dependent nor independent.

Theorem 1.4B. Let V *be an* R-*space in which no nonzero vector is annihilated by* R, *and suppose that* $V = V_1 + \cdots + V_r$, *where each* V_i *is an irreducible subspace of* V. *Then any* r + 1 *vectors in* V *are dependent.*

Proof. Let $X_1, \cdots, X_s$, $s > r$, be any vectors in V. By hypothesis we can write X_i as a sum $\Sigma_j X_{ij}$, where $X_{ij} \in V_j$ for $i = 1, \cdots, s$. We wish to show that $X_1, \cdots, X_s$ are dependent. The theorem is true for r = 0 since all vectors in the 0-space are dependent. Suppose it is true for r - 1. Now if $X_{11} = \cdots X_{s1} = 0$, the induction hypothesis applies. Hence we may suppose, say, that $X_{11} \neq 0$. By hypothesis $RX_{11} \neq 0$. Since V_1 is irreducible, we have by Lemma 1.4A that $RX_{11} = V_1$. In particular, R contains elements $\alpha_{21}, \cdots, \alpha_{s1}$ such that $X_{i1} = \alpha_{i1}X_{11}$. From this point on the proof is just like that of the earlier Theorem 1.3A on vector spaces over a sfield. We subtract $\alpha_{i1}X_1$ from X_i, $i = 2, \cdots, s$, giving s - 1 vectors X_i'

which lie in the space $V' = V_2 + \cdots + V_r$. Since $s > r$, $s - 1 > r - 1$, and so our induction hypothesis asserts that $X_2', \cdots, X_s'$ are dependent. Suppose the vectors so ordered that $X_s' = \alpha_2 X_2' + \cdots + \alpha_{s-1}X_{s-1}'$. Now, substituting $X_i' = X_i - \alpha_{i1}X_1$, we get from this $X_s = (\alpha_{s1} - \alpha_2\alpha_{21} - \cdots - \alpha_{s-1}\alpha_{s-11})X_1 + \alpha_2 X_2 + \cdots + \alpha_{s-1}X_{s-1}$, which establishes the theorem.

Theorem 1.4C. Let $V = W + V_1$, where W and V_1 are subspaces of V, and V_1 is irreducible. Then either $V = W$ or the sum is direct.

Proof. We first show that if $V_1 \cap W \neq 0$, then $W \supseteq V_1$, and hence $V = W$. Suppose that X is a nonzero vector common to W and V_1. If $RX \neq 0$, then by the irreducibility of V_1 we have $RX = V_1$, and hence $V_1 \subseteq W$. If $RX = 0$, then the additive group generated by X is already a nonzero space contained in, and therefore equal to, V_1 and so again $V_1 \subseteq W$. Thus we see that either $V = W$ or $W \cap V_1 = 0$. However, if $W \cap V_1 = 0$ and we express 0 as a sum $0 = X + Y$, where $X \in W$ and $Y \in V_1$, we have $X = -Y$ in both W and V_1 and therefore $X = Y = 0$. This shows that the sum $W + V_1$ is direct.

Corollary 1.4D. Suppose $V = V_1 + \cdots + V_r$, where the V_i are irreducible and none of them can be omitted. Then V is the direct sum of the V_i.

Proof. Let $W_i = V_1 + \cdots + V_i$, $i = 1, \cdots, r$. Now by the above theorem either W_{i+1} is the direct sum of W_i and V_{i+1}, or $V_{i+1} \subseteq W_i$. But if $V_{i+1} \subseteq W_i$, then V_{i+1} could be dropped from the sum for V, contrary to hypothesis. An induction on i shows that each sum $W_i = V_1 + \cdots + V_i$, $i = 1, \cdots, r$, is direct, and the corollary is the case $i = r$.

Before continuing with the theory of completely reducible vector spaces, it is convenient to digress for a while on the theory of isomorphism and homomorphism of spaces. We do this in the next section and then return to the present topic in §6.

.5. ISOMORPHISM AND HOMOMORPHISM OF SPACES

Let V and V' be vector spaces over the same ring R. V is said to be isomorphic (that is, R-isomorphic) to V' if there is a 1-1 mapping $X \leftrightarrow X'$ between the vectors of V and V' which preserves addition and multiplication by elements of R, or, formally, if

(5.1) $$(X + Y)' = X' + Y'$$

and

(5.2) $$(\alpha X)' = \alpha X'.$$

A mapping $X \to X'$ of all the vectors X of V into vectors X' of V' is called a homomorphism (R-homomorphism) if (5.1) and (5.2) are valid. If the image of V is all of V', we speak of a homomorphism of V onto V'. If V' is not necessarily filled by images of V, we speak of a homomorphism of V into V'. If V is mapped homomorphically into V', let W denote the set of all image vectors X'. Then, by (5.1) and (5.2), we see that W is a space onto which V is mapped homomorphically. W is called the homomorphic image of V.

Theorem 1.5A. Let V' be the homomorphic image of an R-space V. Then the set of vectors in V which map into the 0-vector in V' constitutes a subspace V_0 of V. The homomorphism is an isomorphism if and only if $V_0 = 0$.

Proof. Let $X \to X'$ be the given homomorphism of V onto V'. If $X_1' = X_2' = 0$, then by (5.1) $(X_1 + X_2)' = 0$, and by (5.2) $(\alpha X_1)' = \alpha X_1' = 0$. Hence V_0 is a space. Suppose that V_0 contains a nonzero vector X. Then any vector Y' is the image not only of Y but also of $X + Y$, and so the mapping cannot be 1-1 unless $V_0 = 0$. Conversely, if $V_0 = 0$, then no vector Y' can be the image of two different vectors X and Y (for then $(X - Y)' = Y' - Y' = 0$ would imply $V_0 \neq 0$), and so the homomorphism is an isomorphism.

If V is irreducible, we must have either $V_0 = 0$, or $V_0 = V$. This establishes the following fundamental corollary.

Corollary 1.5B. The homomorphic image of an irreducible space is either an isomorphic image or the zero space.

Let V have a subspace W and write $X \equiv Y \pmod{W}$ [read "X is congruent to Y mod W"] if $X - Y \in W$. Denote by $[X]$ the set of all vectors congruent to X (mod W.) We call $[X]$ the residue class of X (mod W). We make the residue classes (mod W) into an R-space by the definitions $[X] + [Y] = [X + Y]$ and $\alpha[X] = [\alpha X]$. To justify these definitions we must show that $X \equiv X'$, $Y \equiv Y'$ imply $X + Y \equiv X' + Y'$ and $X \equiv X'$. But since W is a space $X - X' \in W$ and $Y - Y' \in W$ imply that $(X + Y) - (X' + Y') \in W$

and $\alpha X - \alpha X' = \alpha(X - X') \varepsilon W$. These arguments and definitions establish the following theorem.

Theorem 1.5C. Let W be a subspace of V. Then the residue classes of V (mod W) constitute an R-space homomorphic to V under the mapping $X \rightarrow [X]$, and W is the subspace of V consisting of all vectors whose image is the residue class [0].

This residue class space is also called the factor space of V (mod W) and is denoted by V/W. The importance of factor spaces is apparent from the following converse to Theorem 1.5C.

Corollary 1.5D. Let V' be any homomorphic image of V, and let V_0 be the set of all vectors of V whose image in V' is 0. Then V' is isomorphic to V/V_0 under the mapping $X' \rightarrow [X]$, where $X \rightarrow X'$ is the given homomorphism of V onto V'.

We call V_0 the kernel of the mapping of V onto V'. Then the corollary states that the homomorphic image of a space is isomorphic to the space of residue classes modulo the kernel of the homomorphism.

Proof of the corollary. If $X \equiv Y \pmod{V_0}$, then $(X - Y)' = 0$ or $X' = Y'$. This shows that each residue class X has a unique image X' in V'. Conversely, if $X' = Y'$, then $X - Y \varepsilon V_0$, and so $[X] = [Y]$; hence the mapping $X' \rightarrow [X]$ is indeed 1-1. Finally, we observe that $(X + Y)' = X' + Y'$ and $(\alpha X)' = \alpha X'$, together with $[X + Y] = [X] + [Y]$ and $[\alpha X] = \alpha[X]$, establish the remaining conditions, (5.1) and (5.2), for isomorphism of V' and V/V_0.

6. COMPLETELY REDUCIBLE VECTOR SPACES
(Continuation of §4)

Theorem 1.6A. Let V be the direct sum of the irreducible spaces $V_1, \cdots, V_r$, and let W be any subspace of V. Then W is a direct sum of irreducible subspaces of V.

Proof. We certainly have $V = W + V_1 + \cdots + V_r$. Drop as many V_i's as possible from this relation. We suppose the notation so chosen that this leaves

(6.1) $$V = W + V_{i+1} + \cdots + V_r.$$

If now 0 is written as a sum $0 = Y + X_{i+1} + \cdots + X_r$ of a vector Y in W and vectors $X_{i+1}, \cdots, X_r$ in $V_{i+1}, \cdots, V_r$, respectively, then either $Y = X_{i+1} = \cdots = X_r = 0$ or some $X_j \neq 0$. But, if $X_j = -Y - X_{i+1} - \cdots \neq 0$, we can apply

Theorem 1.4C to $W' = W + V_{i+1} + \cdots + V_{j-1} + V_{j+1} \cdots + V_r$ and V_j to show that V_j could be dropped from the right-hand side of (6.1) without destroying equality, contrary to our hypothesis that no further deletion was possible. This contradiction arose from the premise that the sum (6.1) was not direct; hence the sum (6.1) is direct.

Consider any vector X_1 in V_1. We have $X_1 = Y_1 + X_{i+1} + \cdots + X_r$, where Y_1 is a vector of W uniquely defined by X_1. Clearly $X_1 + X_1' = (Y_1 + Y_1') + (X_{i+1} + X_{i+1}') + \cdots$ and $\alpha X_1 = \alpha Y_1 + \alpha X_{i+1} + \cdots$. Hence the mapping $X_1 \to Y_1$ is a homomorphism of V_1 into W. Let V_1' be the subspace of W onto which V_1 is mapped. If $V_1' = 0$, then $V_1 \subseteq V_{i+1} + \cdots + V_r$, contrary to the hypothesis that V is the direct sum of $V_1, \cdots, V_r$; and therefore by Corollary 1.5B V_1' is ismomorphic to V_1. We obtain in this fashion irreducible subspaces $V_1', \cdots, V_i'$ of W, with V_j' isomorphic to V_j, $j = 1, \cdots, i$. Hence

$$(6.2) \qquad W \supseteq V_1' + \cdots + V_i'.$$

But now $V = V_1' + \cdots + V_i' + V_{i+1} + \cdots + V_r$, since each V_j, $j = 1, \cdots, r$ is contained in the right-hand side. (For $j \leqq i$, $V_j \subseteq V_j' + V_{i+1} + \cdots + V_r$.) Hence any vector Y of W can be written in the form

$$Y = (X_1' + \cdots + X_i') + X_{i+1} + \cdots + X_r.$$

But, since the sum in (6.1) is direct, we conclude from

$$Y = \quad Y \quad + 0 \quad + \cdots + 0$$

that $X_{i+1} = \cdots = X_r = 0$ and $Y = X_1' + \cdots + X_i'$. Hence $W \subseteq V_1' + \cdots + V_i'$. This, with (6.2), gives

$$(6.3) \qquad W = V_1' + \cdots + V_i'.$$

Suppose, now, that some member of the right-hand side of (6.3), say V_1', could be deleted without shrinking W, that is, suppose that $V_1' \subseteq V_2' + \cdots + V_i'$. Then, from $V_1 \subseteq V_1' + V_{i+1} + \cdots + V_r$ and $V_j' \subseteq V_j + V_{i+1} + \cdots + V_r$, $j = 2, \cdots, i$, we would get $V_1 \subseteq V_2 + \cdots + V_r$, contrary to the hypothesis that V is the direct sum of $V_1, \cdots, V_r$. Hence no V_j' can be omitted from (6.3), and now it follows from Corollary 1.4D that W is the direct sum of $V_1', \cdots, V_i'$.

Theorem 1.6B. Let V be the direct sum of irreducible spaces $V_1, \cdots, V_r$ and let W be any irreducible subspace of V. Suppose that W is isomorphic to $V_1, \cdots, V_i$ and not isomorphic to any other V_j. Then $W \subseteq V_1 + \cdots + V_i$.

Proof. We have for $Y \varepsilon W$ a unique expression $Y = X_1 + \cdots + X_r$ with $X_j \varepsilon V_j$. The mapping $Y \rightarrow X_j$ is clearly a homomorphism of W into V_j. Let W_j denote the image of W in V_j. If $W_j \neq 0$, then, since V_j is irreducible and W_j is a space, we have $V_j = W_j$. Furthermore, according to Corollary 1.5B, $W_j \neq 0$ requires that W be isomorphic to W_j and, therefore, in turn, to V_j. Thus, if V_j and W are not isomorphic, $W_j = 0$. In particular, $X_{i+1} = \cdots = X_r = 0$ for every Y in W, which is just what the theorem claims.

Theorem 1.6C. Suppose that $V = V_1 + \cdots + V_r = W_1 + \cdots + W_s$, where each V_j, W_j is irreducible and both sums are direct. Then $r = s$. Moreover, if W_1 is isomorphic to $V_1, \cdots, V_i$ and to no other V_j's and V_1 is isomorphic to $W_1, \cdots, W_k$ and to no other W_j's, then $k = i$ and $V_1 + \cdots + V_i = W_1 + \cdots + W_i$.

Proof. By the theorem just proved $W_j \subseteq V_1 + \cdots + V_i$, $j = 1, \cdots, k$ and $V_j \subseteq W_1 + \cdots + W_k$, $j = 1, \cdots, i$. Hence $V_1 + \cdots + V_i = W_1 + \cdots + W_k$. Now $r = s$ and $i = k$ follow from Theorem 1.4B and the fact that all the sums here are actually direct.

Remarks concerning uniqueness of decomposition into subspaces. If a space V can be written as a direct sum of irreducible subspaces in two ways, we can, by Theorem 1.6C, set up a 1-1 correspondence between the summands, in which corresponding summands are isomorphic. In general, the decomposition into irreducible subspaces is not unique. If, however, we have a direct sum $V = V_{11} + \cdots + V_{1s_1} + V_{21} + \cdots + V_{rs_r}$, where all V_{ij} with the same first subscript are isomorphic and no two V_{ij}'s with different first subscripts are isomorphic, and set $V_i = V_{i1} + \cdots + V_{is_i}$ $i = 1, \cdots, r$, then the sum $V = V_1 + \cdots + V_r$ is direct and this decomposition of V is (again by Theorem 1.6C) unique. Inside V_1 the situation is like that of ordinary vector spaces over a field. For instance, in three space all lines (through the origin) are irreducible isomorphic spaces. Any three noncoplanar lines have as their sum the whole space (and the sum is direct.)

7. RING AS VECTOR SPACE AS WELL AS OPERATOR DOMAIN

We now consider a ring R as an R-vector space, the operation of R (as ring) on R (as space) being the ring multiplication from the left. In other words, the element $\alpha\beta$ of R is interpreted as the vector obtained by applying the operator α to the vector β. The requirements (3.1) on this operation then reduce to certain of the ring postulates. The conditions that a subset $\mathfrak{l}$ of R shall be a subspace are, first, that $\mathfrak{l}$ shall be a subgroup and, second, that $R\mathfrak{l} \subseteq \mathfrak{l}$, that is, precisely the conditions that $\mathfrak{l}$ be a left ideal.

The residue class calculus introduced in §5 for any space applies in particular to the residue classes modulo any left ideal $\mathfrak{l}$. By $R/\mathfrak{l}$ we mean the residue class space. Using brackets to denote residue classes (mod $\mathfrak{l}$) we have, as in §5,

$$[\alpha] + [\beta] = [\alpha + \beta] \quad \text{and} \quad \gamma[\alpha] = [\gamma\alpha].$$

Observe that no multiplication of one residue class by another is involved in dealing with $R/\mathfrak{l}$. If, however, we start with a two-sided ideal $\mathfrak{o}$ instead of a left ideal, we can form $R/\mathfrak{o}$, just as we did $R/\mathfrak{l}$, but then we can make $R/\mathfrak{o}$ into a ring by introducing a multiplication between residue classes. Indeed, we set $[\alpha][\beta] = [\alpha\beta]$. To justify this definition we have only to show that $\alpha \equiv \gamma$, $\beta \equiv \delta \pmod{\mathfrak{o}}$ imply $\alpha\beta \equiv \gamma\delta$ or $\alpha\beta - \gamma\delta \equiv 0 \pmod{\mathfrak{o}}$. We write $\alpha\beta - \gamma\delta$ in the form $\alpha(\beta - \delta) + (\alpha - \gamma)\delta$. Each summand has a factor belonging to the two-sided ideal $\mathfrak{o}$, and so the sum is in $\mathfrak{o}$. We omit the almost trivial verification of the distributive and associative laws. We call the ring thus defined the <u>residue class ring</u> (or, sometimes, <u>difference ring</u>) of R modulo $\mathfrak{o}$ and designate it by $R - \mathfrak{o}$. We reëmphasize that residue class spaces of a ring can be formed modulo any left ideal, but that the residue class ring exists only when the modulus is a two-sided ideal. We also note that $R/\mathfrak{o}$ and $R - \mathfrak{o}$ are not at all the same thing, even though $\mathfrak{o}$ is a two-sided ideal; the first is an R-space and the second is a ring.

We say that R' is a homomorphic image of R if there is a mapping $\alpha \to \alpha'$ of R onto R' such that $(\alpha + \beta)' = \alpha' + \beta'$ and $(\alpha\beta)' = \alpha'\beta'$. The <u>kernel</u> of this mapping is defined to be the set of all elements of R whose image is zero. We state without proof the following analogue of Corollary 1.5D.

<u>Theorem 1.7A. Let R' be a homomorphic image of R and let $\mathfrak{o}$ be the kernel of the mapping. Then $\mathfrak{o}$ is a two-sided ideal and R' is isomorphic to the residue class ring $R - \mathfrak{o}$. Conversely, if $\mathfrak{o}$ is any two-sided ideal of R, then $R - \mathfrak{o}$ is the homomorphic image of R under a mapping which has $\mathfrak{o}$ as kernel.</u>

The three following theorems describe the subsets of R that annihilate vectors and spaces.

Theorem 1.7B. Let V *be any* R*-space. The set* $\mathfrak{d}$ *of all* α *in* R *which annihilate* V *is a two-sided ideal of* R.

Proof. $\mathfrak{d}$ is evidently a subgroup. If now $\alpha V = 0$, we have $(\beta\alpha)V = \beta(\alpha V) = 0$ for any β in R. Furthermore, since V is a space, $\beta V \subseteq V$, and so $(\alpha\beta)V = \alpha(\beta V) \subseteq \alpha V = 0$. Hence $R\mathfrak{d} \subseteq \mathfrak{d}$ and $\mathfrak{d}R \subseteq \mathfrak{d}$.

Theorem 1.7C. Let V *be any* R*-space and* X *any element of it. The set* $\mathfrak{l}$ *of all* λ *in* R *which annihilate* X *is a left ideal of* R.

Proof. If $\lambda X = 0$, then $(\alpha\lambda)X = \alpha(\lambda X) = 0$ for all $\alpha \in R$. Furthermore, if $\lambda_1 X = \lambda_2 X = 0$, then $(\lambda_1 - \lambda_2)X = \lambda_1 X - \lambda_2 X = 0 - 0 = 0$, so that $\mathfrak{l}$ is a subgroup. Hence $\mathfrak{l}$ is a left ideal.

Theorem 1.7D. Let $V = RX$ *be an irreducible vector space. Then the set* $\mathfrak{l}$ *of annihilators of* X *is a maximal left ideal of* R *and* V *is isomorphic to the space* $R/\mathfrak{l}$.

Proof. Any vector Y in V is of the form βX ($\beta \in R$). The relations $\beta_1 X + \beta_2 X = (\beta_1 + \beta_2)X$ and $(\alpha\beta)X = \alpha(\beta X)$ show that the mapping $\beta \to Y$ is a homomorphism of R (as space) onto V. By the preceding theorem the kernel of this homomorphism is a left ideal, $\mathfrak{l}$, of R, and V is isomorphic to $R/\mathfrak{l}$ by Corollary 1.5D. If $\mathfrak{l}'$ is any left ideal of R, it is clear that $\mathfrak{l}'X$ is a subspace of V. If, in particular, $\mathfrak{l}' \supset \mathfrak{l}$, then $\mathfrak{l}'X \neq 0$; hence, by the irreducibility of V, $\mathfrak{l}'X = V$. Now apply Theorem 1.7C and Corollary 1.5D to V considered as an $\mathfrak{l}'$-space; $\mathfrak{l}$ is still the annihilator of X, and so we get $\mathfrak{l}'/\mathfrak{l}$ isomorphic to V and, therefore, in turn, to $R/\mathfrak{l}$. This final isomorphism between $\mathfrak{l}'/\mathfrak{l}$ and $R/\mathfrak{l}$ is actually obtained by identifying the residue classes of $\mathfrak{l}'$ (mod $\mathfrak{l}$) with those of R (mod $\mathfrak{l}$); in other words, R and $\mathfrak{l}'$ are identical. This shows that $\mathfrak{l}$ is maximal in R.

CHAPTER II

MINIMUM CONDITION

1. MINIMUM CONDITION

Let R be a ring, and V a vector space with R as left operator domain. V is said to have minimum condition on subspaces (that is, on R-subspaces) if every set of subspaces has in it a subspace which contains no other member of the set. We shall call such a subspace minimal. A minimal subspace of the set of all subspaces of V is called a minimal subspace of V. The minimum condition implies that every descending chain of subspaces of V,

$$V \supset V_1 \supset V_2, \cdots \tag{1.1}$$

has a minimal member, and hence must terminate. Conversely, if every descending chain of subspaces of V terminates, then V has the minimum condition on subspaces. For let S be a set of subspaces of V, then every descending chain of subspaces from the set S must terminate. If V_1 is a member of S, then either V_1 is minimal, and we are finished, or $V_1 \supset V_2$, where V_2 is another subspace of S. Then V_2 is either minimal, or $V_2 \supset V_3$, etc. In this manner we obtain a descending chain of subspaces of V which terminates with a minimal subspace.

We say that R has minimum condition on left ideals if R considered as a left R-space has minimum condition on subspaces. The left R-subspaces of R are just the left ideals. Thus the concept of minimum condition on subspaces is more general than that of minimum condition on left ideals.

It was formerly found expedient to impose a maximum as well as a minimum condition on ideals. However, a simple proof which requires only the hypothesis of the minimum condition on left ideals has now been given by R. Brauer* for the important theorem that a nonnilpotent left ideal contains an idempotent element (cf. Theorem 2.4A).

---

*See R. Brauer, "On the nilpotency of the radical of a ring," Bull. Amer. Math. Soc., 48 (1942), 752-758.

When we refer to "minimum condition in a ring or space" we shall always understand "minimum condition on left ideals" or "minimum condition on subspaces," respectively. In the following argument we shall always assume, unless otherwise stated, the minimum condition on left ideals

2. MINIMUM CONDITION IN SPACES

In the following discussion we suppose the minimum condition checked in R and ask for conditions under which it will hold in an R-space V.

Lemma 2.2A. *Let* R *be any ring and let* $V = W + U$, *where* V, W, U *are (left)* R-*spaces, of which* W *and* U *have minimum condition on subspaces. Then the minimum condition holds in* V.

Proof. Let V' be a subspace of V, and suppose $X' = * + Z' \varepsilon V'$, where $Z' \varepsilon U$ and $* \varepsilon W$. For a given X' there may be several possibilities for Z'. Let U' be the set of all Z' obtained as X' runs through V'; and let $W' = V' \cap W$. Since $\alpha X' = \alpha * + \alpha Z'$, U' is an R-space, and W' as the intersection of two spaces is again a space. We say that U' and W' are *related* to V'.

It is evident that if $V'' \subseteq V'$, then $U'' \subseteq U'$, and $W'' \subseteq W'$, that is, the relations between V' and U', V' and W' are monotone.

The spaces U' and W' are uniquely defined by V'. We might ask if, conversely, U' and W' can come from more than one V subspace. The following example shows that the answer is yes. Let U be any space and let the mapping $Z \rightarrow \sigma(Z)$ be an isomorphism of U onto a second space W. Take for V' the set of all vectors $Z + \sigma(Z)$, and for V'' the set of all vectors $Z - \sigma(Z)$. Then $U' = U'' = U$, and $W' = W'' = 0$, although $V' \neq V''$. However, we can assert that $V'' \subseteq V'$, $U'' = U'$, $W'' = W' \Rightarrow V'' = V'$. For suppose $X' = * + Z' \varepsilon V'$. Then, since $U'' = U'$, there is an X'' in V'' such that $X'' = *_1 + Z'$. Now $X' - X'' \varepsilon V'$ and $* - *_1 \varepsilon W$; since the two are equal they both belong to W'. Hence $X' = X'' + (* - *_1) \varepsilon V'' + W' = V'' + W'' = V''$, or $V' \subseteq V''$. This proves that $V' = V''$.

We now complete the proof of the lemma. Let $S(V)$ be any set of subspaces of V, and denote by $S(U)$ the set of related subspaces of U. Choose U' minimal in $S(U)$. Those elements (that is, subspaces of V) of $S(V)$ to which U' is related constitute a subset $S'(V)$ of $S(V)$. Denote by $S'(W)$ the set of subspaces of W related to the elements of $S'(V)$; choose W' minimal in $S'(W)$; and let V' be any element of $S'(V)$ related to W'. Then V' is minimal in $S(V)$. For suppose $V'' \subseteq V'$. By the monotone properties of the relations, we have then $U'' \subseteq U'$, $W'' \subseteq W'$. But, by the minimal nature of U', this requires $U' = U''$, and by the minimal nature of W', $W' = W''$. But we have just seen that these equalities together with $V'' \subseteq V'$ imply $V' = V''$.

The following corollary comes from the lemma by an induction argument.

Corollary 2.2B. Let $V_1, \cdots, V_m$ _be R-spaces satisfying the minimum condition. Then_ $V = V_1 + \cdots + V_m$ _satisfies the minimum condition._

Theorem 2.2C. Let R _be a ring with minimum condition, and let the R-space_ V _be covered by a finite sum_ $RX_1 + \cdots + RX_m$. _Then_ V _satisfies the minimum condition._

Proof. We first note that RX is a space, homomorphic to R, and so satisfies the minimum condition. Then apply Corollary 2.2B.

Theorem 2.2D. Let R _be a ring with a subring_ R' _such that_ $R = R_1 + \cdots + R_m$, _where each_ R_i _is an_ R'_-space satisfying the minimum condition. Then the minimum condition holds in_ R.

Proof. Applying Corollary 2.2B, we obtain that R, considered as an R'-space, satisfies the minimum condition. Every left ideal of R is certainly an R'-space, and so R as a ring satisfies the minimum condition.

Theorem 2.2E. Let R _be a ring with minimum condition. The set_ S _of all_ m_-rowed square matrices with elements from_ R _is a ring with minimum condition._

Proof. We identify the element α of R with the diagonal matrix $\begin{Vmatrix} \alpha & & \\ & \ddots & \\ & & \alpha \end{Vmatrix}$ and so have R as subring of S. Let R_{ik} denote the set of all matrices with zeros everywhere except for elements of R in the i-th row and k-th column intersection. R_{ik} is an R-space homomorphic to R and so satisfies the minimum condition. Consider now $S = \Sigma R_{ik}$ as a ring containing the subring R. Then from Theorem 2.2D it follows that the minimum condition holds in S.

3. NILPOTENT IDEALS

An element α of a ring R is called nilpotent if a power $\alpha^m = 0$. We say that an ideal $\mathfrak{a}$ is nilpotent if a power $\mathfrak{a}^m$ is the null ideal (0). In this section we discuss briefly nilpotent ideals, and here no hypothesis of minimum condition holding in R is necessary.

Theorem 2.3A. The sum $\mathfrak{l}_1 + \mathfrak{l}_2$ of two nilpotent left ideals $\mathfrak{l}_1$, $\mathfrak{l}_2$ is a nilpotent left ideal.

Proof. Let $\mathfrak{l}_1^n = 0$, $\mathfrak{l}_2^m = 0$. Each term of $(\mathfrak{l}_1 + \mathfrak{l}_2)^{n+m}$ will contain either $p \geq n$ factors $\mathfrak{l}_1$ or $q \geq m$ factors $\mathfrak{l}_2$. A term ρ which contains $p \geq n$ factors $\mathfrak{l}_1$ is of form

$$\cdots \mathfrak{l}_1 \cdots \mathfrak{l}_1 \cdots \mathfrak{l}_1 \cdots \mathfrak{l}_1 \cdots,$$

where the dots denote factors $\mathfrak{l}_2$ and, since $A\mathfrak{l}_1 \subseteq \mathfrak{l}_1$ for any subset A of R, $\rho \subseteq \mathfrak{l}_1^p \cdots = (0)$. It follows that $(\mathfrak{l}_1 + \mathfrak{l}_2)^{n+m} = (0)$, so that $\mathfrak{l}_1 + \mathfrak{l}_2$ is nilpotent.

Theorem 2.3B. If $\mathfrak{l}$ is a nilpotent left ideal, then $\mathfrak{l}R$ is a nilpotent two-sided ideal.

Proof. Let $\mathfrak{l}^n = (0)$. Then $(\mathfrak{l}R)^n$ may be written as

$$\mathfrak{l} \cdot R\mathfrak{l} \cdot R\mathfrak{l} \cdots R\mathfrak{l} \cdot R \subseteq \mathfrak{l}^n \cdot R = (0).$$

Corresponding theorems hold for right ideals, namely:

Theorem 2.3A'. The sum $\mathfrak{r}_1 + \mathfrak{r}_2$ of two nilpotent right ideals $\mathfrak{r}_1$, $\mathfrak{r}_2$ is a nilpotent right ideal.

Theorem 2.3B'. If $\mathfrak{r}$ is a nilpotent right ideal, then $R\mathfrak{r}$ is a nilpotent two-sided ideal.

4. IDEMPOTENT ELEMENTS

An element α of a ring R is called an idempotent if $\alpha^2 = \alpha$. An idempotent $\alpha \neq 0$ is evidently not nilpotent. Let us exclude the zero idempotent from our consideration, so that in the following argument when we speak of an idempotent we shall always mean an idempotent other than zero. We now use our assumption that R has the minimum condition on left ideals to prove

Theorem 2.4A. If the left ideal $\mathfrak{l}$ of R is nonnilpotent, then $\mathfrak{l}$ contains an idempotent.

Proof. From the set of nonnilpotent subideals of $\mathfrak{l}$ we select a minimal subideal $\mathfrak{l}_1$. Then $\mathfrak{l}_1^2 = \mathfrak{l}_1$, since

$\mathfrak{l}_1^2 \subseteq \mathfrak{l}_1$ and $\mathfrak{l}_1^2$ cannot be nilpotent because $\mathfrak{l}_1$ is nonnilpotent. We look now for the set M of left ideals $\mathfrak{m}$ such that

$$(1)\ \mathfrak{l}_1\mathfrak{m} \neq 0 \qquad (2)\ \mathfrak{m} \subseteq \mathfrak{l}_1.$$

Since $\mathfrak{l}_1$ itself satisfies (1), (2), such ideals $\mathfrak{m}$ exist. From the set M we select a minimal ideal $\mathfrak{m}_1$. By (1) there is an element $\mu \neq 0$ of $\mathfrak{m}_1$ such that $\mathfrak{l}_1\mu \neq 0$. We contend that $\mathfrak{l}_1\mu = \mathfrak{m}_1$. First, $\mathfrak{l}_1$ is a left ideal (cf. §1.2). Further, $\mathfrak{l}_1\mu \subseteq \mathfrak{m}_1 \subseteq \mathfrak{l}_1$, and $\mathfrak{l}_1 \cdot \mathfrak{l}_1\mu = \mathfrak{l}_1^2\mu = \mathfrak{l}_1\mu \neq 0$, so that $\mathfrak{l}_1\mu$ is an ideal of the set M, and is contained in $\mathfrak{m}_1$. As $\mathfrak{m}_1$ is minimal, we have then $\mathfrak{l}_1\mu = \mathfrak{m}_1$.

It now follows that, since $\mu \in \mathfrak{m}_1$, there exists an element λ in $\mathfrak{l}_1$ such that $\lambda\mu = \mu$. But then $\mu = \lambda^n\mu$ for all integers n, and as $\mu \neq 0$, λ cannot be nilpotent. Thus we have proved that $\mathfrak{l}_1$ contains a nonnilpotent element.

Further, we have that $\lambda^2\mu = \lambda\mu$, or $(\lambda^2 - \lambda)\mu = 0$. The set of elements τ in $\mathfrak{l}_1$ for which $\tau\mu = 0$ is a left ideal $\mathfrak{n}$ which is contained in $\mathfrak{l}_1$ (cf. Theorem 1.7C), and our relation shows $\lambda^2 - \lambda \in \mathfrak{n}$. Since $\mathfrak{n}\mu = 0$ and $\mathfrak{l}_1\mu \neq 0$, $\mathfrak{n}$ cannot be the whole ideal $\mathfrak{l}_1$. Then $\mathfrak{n}$ is a proper subideal of $\mathfrak{l}_1$, and is consequently nilpotent.

We now have that λ is not nilpotent but that $\lambda^2 - \lambda = x$ is nilpotent, say, $x^n = 0$. We form $\lambda_1 = \lambda + x - 2\lambda x$ and shall show that λ_1 is not nilpotent, but $\lambda_1^2 - \lambda_1$ is nilpotent. We observe λ, x, λ_1 are commutative with each other with respect to multiplication. The element λ_1 is not nilpotent, for if λ_1 were nilpotent, then $\lambda = \lambda_1 - x + 2\lambda x$, as sum of terms which are commutative and nilpotent, would also be nilpotent. A calculation shows, on making use of the relation $\lambda^2 = \lambda + x$, that $\lambda_1^2 - \lambda_1 = 4x^3 - 3x^2 = x_1$, say. Here $\lambda_1^2 - \lambda_1$ is expressed by higher powers of the nilpotent element x than is $\lambda^2 - \lambda = x$. Repeating the process by setting $\lambda_2 = \lambda_1 + x - 2\lambda_1x_1$, we obtain still higher powers of x in the expression $\lambda_2^2 - \lambda_2$, and continuing we obtain finally a not nilpotent element λ_r with $\lambda_r^2 - \lambda_r = 0$. Thus refinement of the not nilpotent elements $\lambda, \lambda_1, \cdots$ produces an idempotent element of $\mathfrak{l}$.

We now have a criterion for a left ideal being nilpotent, namely,

Corollary 2.4B. A necessary and sufficient condition that a left ideal $\mathfrak{l}$ be nilpotent is that every element of $\mathfrak{l}$ be nilpotent.

The necessity of this condition follows from the definition of a nilpotent ideal, while that the condition is sufficient is a direct consequence of Theorem 2.4A.

Corollary 2.4C. Any sum $\mathfrak{l} = \mathfrak{l}_1 + \mathfrak{l}_2 \cdots$ of nilpotent left ideals is a nilpotent left ideal.

Proof. That $\mathfrak{l}$ is a left ideal follows by Theorem 1.2A. Further, each element α of the sum is contained in a finite sum of the nilpotent ideals $\mathfrak{l}_i$. We obtain immediately from Theorem 2.3A that a finite sum of nilpotent ideals is a nilpotent ideal. Then each element α of the sum $\mathfrak{l}$ is nilpotent, and by Corollary 2.4B, $\mathfrak{l}$ is a nilpotent ideal.

5. THE RADICAL OF A RING

The radical N of a ring R we define as the sum of all nilpotent left ideals of R.

Theorem 2.5A. The radical N of the ring R is a nilpotent two-sided ideal, and contains all nilpotent right ideals of R as well as all nilpotent left ideals of R.

Proof. By Corollary 2.4C, N is nilpotent; in fact, N is the largest of all nilpotent left ideals of R. Then N contains all nilpotent right ideals. For let $\mathfrak{r}$ be a nilpotent right ideal. By Theorem 2.3B', $R\mathfrak{r}$ is a nilpotent two-sided ideal. Consequently, from Theorem 2.3A', $\mathfrak{r} + R\mathfrak{r}$ is nilpotent, and it is easily verified that $\mathfrak{r} + R\mathfrak{r}$ is a left ideal. Then $\mathfrak{r} + R\mathfrak{r} \subseteq N$, and so $\mathfrak{r} \subseteq N$.

Since NR is by Theorem 2.3B a nilpotent left ideal, then $NR \subseteq N$, that is, N is a two-sided ideal.

6. PEIRCE DECOMPOSITION

To study the structure of nonnilpotent left ideals we prove

Theorem 2.6A. Let $\mathfrak{l}$ be any nonnilpotent left ideal of R, let $e_{\alpha_1}, e_{\alpha_2}, \cdots$ denote idempotent elements in $\mathfrak{l}$, and $\mathfrak{m}_{e_\alpha}$ the left subideal of $\mathfrak{l}$ of elements of $\mathfrak{l}$ which are left annihilators of e_α. If e be such an idempotent that $\mathfrak{m}_e$ is a minimal ideal of the set of ideals $\mathfrak{m}_{e_\alpha}$, then $\mathfrak{m}_e$ is nilpotent.

Proof. If $\mathfrak{m}_e$ were not nilpotent, then $\mathfrak{m}_e$ contains an idempotent element e_1, and $e_1 e = 0$. We form $e' = e + e_1 - ee_1$ and observe that $e'e = e$, $ee' = e$, $e_1 e' = e_1$,

$e'^2 = (e + e_1 - ee_1)e' = e + e_1 - ee_1 = e'$. Now consider the subideal $\mathfrak{m}_{e'}$ of elements of $\mathfrak{l}$ which are left annihilators of e'. If $xe' = 0$, then $xe'e = xe = 0$, so that $\mathfrak{m}_{e'} \subseteq \mathfrak{m}_e$. But $e_1e' = e_1 \neq 0$ while $e_1e = 0$. This implies $\mathfrak{m}_{e'}$ is a proper subideal of $\mathfrak{m}_e$, contrary to our assumption that $\mathfrak{m}_e$ is minimal.

Theorem 2.6B. A nonnilpotent left ideal $\mathfrak{l}$ of R may be decomposed into a direct sum $\mathfrak{l} = \mathfrak{l}e + \mathfrak{m}$, where $\mathfrak{m}$ is nilpotent.

Proof. We choose e as in Theorem 2.6A, and write any element x of $\mathfrak{l}$ as

(6.1) $x = xe + (x - xe)$ (Peirce decomposition).

Since $(x - xe)e = 0$, $x - xe \in \mathfrak{m}_e$. On the other hand, any element $x \in \mathfrak{m}_e$ may be written as $x - xe$, since here $x \cdot e = 0$; consequently, the set of all elements $x - xe$, $x \in \mathfrak{l}$ is just $\mathfrak{m}_e$.

The splitting (6.1) of x is unique, and, consequently, we have on setting $\mathfrak{m}_e = \mathfrak{m}$ that $\mathfrak{l} = \mathfrak{l}e + \mathfrak{m}$ is a direct sum.

Remarks. The decomposition of $\mathfrak{l}$ may also be written as a direct sum $\mathfrak{l} = Re + \mathfrak{m}$. For $Re \subseteq \mathfrak{l}$, and then $Re \cdot e = Re \subseteq \mathfrak{l}e$, but $\mathfrak{l}e \subseteq Re$, so $Re = \mathfrak{l}e$. A nonnilpotent left ideal $\mathfrak{l}$ of R is then almost a principal ideal with generating element e.

We may apply Theorem 2.6B to R itself and obtain a direct decomposition $R = Re + \mathfrak{m}$, where $\mathfrak{m}$ is nilpotent. In case R contains a unit element e, this relation contracts to $R = Re$.

Theorem 2.6C. Let $\mathfrak{a}$ be a nonnilpotent two-sided ideal. Then an idempotent element e in $\mathfrak{a}$ may be chosen so that $\mathfrak{a}$ may be written both as

$$\mathfrak{a} = \mathfrak{a}e + \mathfrak{m}$$
$$\mathfrak{a} = e\mathfrak{a} + \mathfrak{n} \tag{6.2}$$

where $\mathfrak{m}$, $\mathfrak{n}$ are nilpotent, and both sums are direct.

Considering $\mathfrak{a}$ as a left ideal, we choose e as in Theorem 2.6A, so that by Theorem 2.6B $\mathfrak{a} = \mathfrak{a}e + \mathfrak{m}$ (direct sum) with $\mathfrak{m}$ nilpotent. We obtain a direct decomposition $\mathfrak{a} = e\mathfrak{a} + \mathfrak{n}$ by taking in $\mathfrak{n}$ all elements $x - ex$, x any element of $\mathfrak{a}$, so that $\mathfrak{n}$ is the right ideal of elements of $\mathfrak{a}$ which are right annihilators of e. We shall now show that $\mathfrak{n}$ is nilpotent.

From the direct splitting $\mathfrak{a} = \mathfrak{a}e + \mathfrak{m}$ we obtain, since $\mathfrak{n} \subseteq \mathfrak{a}$, that $\mathfrak{n} = \mathfrak{n}e + \mathfrak{m}_0$ (direct sum), where $\mathfrak{m}_0 \subseteq \mathfrak{m}$. It follows that $\mathfrak{n}^2 = (\mathfrak{n}e + \mathfrak{m}_0)\mathfrak{n} = \mathfrak{m}_0\mathfrak{n}$ since $e\mathfrak{n} = 0$. Then $\mathfrak{n}^3 = \mathfrak{m}_0\mathfrak{n}^2 = \mathfrak{m}_0^2\mathfrak{n}$, $\mathfrak{n}^{\rho+1} = \mathfrak{m}_0^{\rho}\mathfrak{n}$, and since $\mathfrak{m}_0$ is part of the nilpotent ideal $\mathfrak{m}$, $\mathfrak{n}$ must also be nilpotent.

Theorem 2.6D. A nonnilpotent two-sided ideal $\mathfrak{a}$ *may be decomposed into a direct sum* $\mathfrak{a} = \mathfrak{a}_{11} + \mathfrak{a}_{10} + \mathfrak{a}_{01} + \mathfrak{a}_{00}$, *where the ring* $\mathfrak{a}_{11}$ *has* e *as left and right unit,* $\mathfrak{a}_{10}$ *has* e *as left unit and right annihilator,* $\mathfrak{a}_{01}$ *has* e *as right unit and left annihilator,* $\mathfrak{a}_{00}$ *has* e *as left and right annihilator, and the rings* $\mathfrak{a}_{10}$, $\mathfrak{a}_{01}$, $\mathfrak{a}_{00}$ *are all nilpotent.*

Proof. We select e as in 2.3A, and for each element x of $\mathfrak{a}$ make the unique splitting

$$(6.3) \qquad x = exe + (ex - exe) + (xe - exe) + (x + exe - ex - xe).$$

The elements exe comprise the ring $\mathfrak{a}_{11}$, the elements $ex - exe$ the ring $\mathfrak{a}_{10}$, etc. In the decomposition (6.2) of $\mathfrak{a}$, $\mathfrak{a}_{10} + \mathfrak{a}_{00}$ gives $\mathfrak{m}$, and $\mathfrak{a}_{01} + \mathfrak{a}_{00}$ gives $\mathfrak{n}$.

CHAPTER III

MATRIX REPRESENTATIONS

1. MATRICES

Let $\mathfrak{k}$ be a sfield. By an $m \times n$ matrix ($\mathfrak{k}$-matrix) we mean a rectangular array of m rows, n columns

$$(1.1) \qquad A = \begin{Vmatrix} a_{11} & a_{12} & \cdots & a_{1n} \\ a_{21} & a_{22} & \cdots & a_{2n} \\ \cdot & \cdot & \cdots & \cdot \\ a_{m1} & a_{m2} & \cdots & a_{mn} \end{Vmatrix}$$

with elements a_{ik} in the sfield $\mathfrak{k}$. We refer to m and n as the <u>dimensions</u> of the matrix. We adopt the notation

$$A = \| a_{ik} \|,$$

where i is the row index, k the column index.

Consider, now, the set of all matrices with coefficients in the sfield $\mathfrak{k}$. We can add two matrices if they have the same dimensions, using the rule

$$(1.2) \qquad \| a_{ik} \| + \| b_{ik} \| = \| a_{ik} + b_{ik} \| .$$

We can multiply two matrices if they have the same interior dimension,

$$(1.3) \qquad \| a_{ik} \| \cdot \| b_{ik} \| = \| a_{i\nu} b_{\nu k} \| ,$$

where the repeated ν indicates summation over the range determined by the interior dimension. We shall now prove that the associative and distributive laws hold for matrix products and sums where these exist. To prove the associative law for products we observe that

$$(\| a_{ik} \| \cdot \| b_{ik} \|) \| c_{ik} \| = \| a_{i\nu} b_{\nu k} \| \cdot \| c_{ik} \| = \| a_{i\nu} b_{\nu\mu} c_{\mu k} \|$$

and

$$\| a_{ik} \| (\| b_{ik} \| \cdot \| c_{ik} \|) = \| a_{ik} \| \cdot \| b_{i\mu} c_{\mu k} \| = \| a_{i\nu} b_{\nu\mu} c_{\mu k} \| .$$

Distributivity follows easily, for

$$\| a_{ik} \| (\| b_{ik} + c_{ik} \|) = \| a_{ik} \| \cdot \| b_{ik} + c_{ik} \|$$
$$= \| a_{i\nu} b_{\nu k} + a_{i\nu} c_{\nu k} \| = \| a_{i\nu} b_{\nu k} \| + \| a_{i\nu} c_{\nu k} \| = \| a_{ik} \| \cdot \| b_{ik} \|$$
$$+ \| a_{ik} \| \cdot \| c_{ik} \| ,$$

and a similar computation holds for $(\|b_{ik}\| + \|c_{ik}\|)\|a_{ik}\|$.

We observe that among the matrices with elements in the sfield $\mathfrak{k}$ there are the $1 \times n$ matrices or row vectors of form $a = (a_1, \cdots, a_n) = (a_k)$, and the $m \times 1$ matrices,

or column vectors of form $a = \begin{Vmatrix} a_1 \\ a_2 \\ \cdot \\ \cdot \\ \cdot \\ a_m \end{Vmatrix} = (a_i)$.

We shall now prove

<u>Theorem 3.1A</u>. <u>Let</u> V_m, V_n <u>be</u> $\mathfrak{k}$-<u>right spaces of column vectors of dimension</u> m, n <u>respectively, and let coördinate systems</u> $(X_1, \cdots, X_m)$, $(Y_1, \cdots, Y_n)$ <u>be chosen for</u> V_m, V_n. <u>Then, with regard to those coördinate systems, an</u> $n \times m$ <u>matrix</u> $A = \|a_{ik}\|$ <u>defines a</u> $\mathfrak{k}$-<u>homomorphic mapping of</u> V_m <u>into</u> V_n, <u>and, conversely, a</u> $\mathfrak{k}$-<u>homomorphism</u> σ <u>of</u> V_m <u>into</u> V_n <u>determines an</u> $n \times m$ <u>matrix</u> A_σ <u>with elements in</u> $\mathfrak{k}$.

<u>Proof</u>. Any vector X of V_m may be written

$$X = X_1 c_1 + \cdots + X_m c_m = (X_k)(c_i), \tag{1.4}$$

with elements c_i from $\mathfrak{k}$, and, similarly, any vector Y of V_n has the form

$$Y = Y_1 d_1 + \cdots + Y_n d_n = (Y_k)(d_i). \tag{1.5}$$

From the relations

$$\begin{aligned} X_k &\rightarrow Y_\nu a_{\nu k} \\ (X_k) &\rightarrow (Y_k)\|a_{ik}\| \\ (X_k)(c_i) &\rightarrow (Y_k)\|a_{ik}\|(c_i) = (Y_k)(d_i), \end{aligned} \tag{1.6}$$

where $d_i = a_{i\mu} c_\mu$, we obtain from the matrix A a mapping of V_m into V_n, and it may easily be verified that this mapping is a $\mathfrak{k}$-homomorphism.

Conversely, if $X \rightarrow \sigma(X)$ is a $\mathfrak{k}$-homomorphism of V_m into V_n, then we have

$$\begin{aligned} X_k &\rightarrow \sigma(X_k) = Y_\mu a^\sigma_{\mu k} \\ (X_k) &\rightarrow (Y_k)\|a^\sigma_{ik}\|, \end{aligned} \tag{1.7}$$

which leads to the $n \times m$ matrix.

$$A_\sigma = \|a^\sigma_{ik}\|. \tag{1.8}$$

Let A be an $m \times m$ matrix. A matrix A^{-1} is called the inverse of A if $A^{-1}A = AA^{-1} = E_m$, where $E_m = \| \delta_{ik} \|$, $\delta_{ik} = 0$, $i \neq k$, $\delta_{ii} = 1$ $(i, k = 1, 2, \cdots, m)$. The $m \times m$ matrices which possess inverses form a multiplicative group with E_m as unit element.

Corollary 3.1B. Let V_m be an m-dimensional $\mathfrak{k}$-space with coördinate system (X_k). With regard to this coördinate system a $\mathfrak{k}$-isomorphism σ of V_m onto itself determines an $m \times m$ matrix A_σ which possesses an inverse, and, conversely, an $m \times m$ matrix A which possesses an inverse defines a $\mathfrak{k}$-isomorphism of V_m onto itself.

Proof. Let $X \rightarrow \sigma(X)$ be a $\mathfrak{k}$-isomorphism of V_m onto itself, and let $\sigma(X_p) = X_\nu a_{\nu p}$. The mapping $\sigma(X) \rightarrow X$ is also a $\mathfrak{k}$-isomorphism of V_m onto itself; let $B = \| b_{ik} \|$ be the matrix determined by this isomorphism. Then $X_\nu \rightarrow X_\mu b_{\mu\nu}$, and $X_\nu a_{\nu p} \rightarrow X_\mu b_{\mu\nu} a_{\nu p}$. But $X_\nu a_{\nu p} = \sigma(X_p) \rightarrow X_p$, so that $X_\mu b_{\mu\nu} a_{\nu p} = X_p$, that is, $b_{\mu\nu} a_{\nu p} = 0$ $\mu \neq p$, $b_{p\nu} a_{\nu p} = 1$. Then the $m \times m$ matrix $A = \| a_{ik} \|$ determined by the $\mathfrak{k}$-isomorphism σ has B as its left inverse, $BA = E_m$. Similarly, starting with the isomorphism $\sigma(X) \rightarrow X$, we obtain $AB = E_m$, that is, $B = A^{-1}$.

For the converse, we observe that, if $X = (X_k)(c_i) \rightarrow X' = (X_k)A(c_i) = 0$, then $A(c_i) = 0$, since the X_k are linearly independent. Now operating with A^{-1} we have $A^{-1}A(c_i) = (c_i) = 0$, that is, $X = 0$, so that the mapping $X \rightarrow X'$ is (1-1).

Remarks. If in Theorem 3.1A the coördinate system (X_k) is replaced by $(\overline{X}_k) = (X_k)T$, T an $m \times m$ matrix with inverse, and the coördinate system (Y_k) is replaced by $(\overline{Y}_k)$ such that $(Y_k) = (\overline{Y}_k)S$, S an $n \times n$ matrix with inverse, then the matrix A_σ determined by σ is replaced by $SA_\sigma T$. In particular, if $V_m = V_n$ and $(\overline{Y}_k) = (\overline{X}_k)$, $(Y_k) = (X_k)$, then $S = T^{-1}$ and the matrix $T^{-1}A_\sigma T$ corresponds to σ.

2. REPRESENTATION SPACES

Let R be a ring and V an R-left space. We suppose that V is also a $\mathfrak{k}$-right space, $\mathfrak{k}$ a sfield, and that V has dimension n over $\mathfrak{k}$, and that for $\alpha \in R$, $X \in V$, $a \in k$ we have the associativity relation.

$$(2.1) \qquad \alpha(Xa) = (\alpha X)a.$$

Let, now, (X_k) be a $\mathfrak{k}$-basis for V. If α is any element of R, then from (2.1) it follows that

$$(2.2) \qquad X \to X' = \alpha \cdot X$$

is a $\mathfrak{k}$-homomorphism of V into itself. From Theorem 3.1A we have that the mapping (2.2) determines a matrix A_α, with elements from $\mathfrak{k}$ such that

$$(2.3) \qquad \alpha(X_k) = (X_k)A_\alpha.$$

This formula shows us that the operation of combining a ring element with a vector of V can be replaced by matrix multiplication. We shall see that these matrices A_α have very agreeable properties. For if α, $\beta \varepsilon$ R, then from (2.3) and

$$(2.4) \qquad \beta(X_k) = (X_k)A_\beta$$

it follows that

$$(\alpha + \beta)(X_k) = (X_k)(A_\alpha + A_\beta).$$

$(\alpha + \beta)$ is an element of R, so that $(\alpha + \beta)(X_k) = (X_k)A_{\alpha+\beta}$, and then

$$(2.5) \qquad (X_k)(A_\alpha + A_\beta) = (X_k)A_{\alpha+\beta}.$$

Before concluding $A_\alpha + A_\beta = A_{\alpha+\beta}$ we must first establish cancellation. This follows easily, for if $(X_k)C = 0$, $C = \|c_{ik}\|$, then $X_\mu c_{\mu k} = 0$ $(k = 1, 2, \cdots, n)$, and since the X_μ are linearly independent, $c_{ik} = 0$ for all i, k, which is the definition of $C = 0$. Then $(X_k)C = (X_k)D$ implies $(X_k)(C - D) = 0$, that is, $C - D = 0$ or $C = D$. (2.5) then gives

$$(2.6) \qquad A_{\alpha+\beta} = A_\alpha + A_\beta.$$

Furthermore, from

$$\alpha\beta(X_k) = \alpha(\beta(X_k)) = \alpha(X_k)A_\beta = (X_k)A_\alpha A_\beta$$

we obtain

$$(2.7) \qquad A_{\alpha\beta} = A_\alpha \cdot A_\beta.$$

Thus $\alpha \to A_\alpha$ is a homomorphic mapping of the ring R on the ring of matrices A_α. We have proved

<u>Theorem 3.2A</u>. <u>If V is an R-left space and a $\mathfrak{k}$-right space whose elements satisfy the associativity relation (2.1), then a basis (X_k) of V determines by the relation $\alpha(X_k) = (X_k)A_\alpha$ a ring of matrices A_α upon which R is homomorphically mapped.</u>

Any mapping $\alpha \to A_\alpha$ of R into a ring $\mathbf{A}$ of square matrices which satisfy (2.6) and (2.7) is called a <u>representation</u> of R. In the discussion above the matrices A_α

are realizations of the $\mathfrak{k}$-homomorphic mappings of V into itself produced by the elements α of R operating on V. Then V is called a <u>representation space</u> of A.

If another basis $(\overline{X}_k)$ is used, $(\overline{X}_k) = (X_k)T$, $(X_k) = (\overline{X}_k)T^{-1}$, then the resulting representation consists of the matrices $T^{-1}A_\alpha T$, and is said to be <u>equivalent</u> to the representation by the matrices A_α. Equivalent representations are usually considered not essentially different. We shall adopt this viewpoint, and shall refer to the representation given by a representation space.

<u>Theorem 3.2B. Let</u> A <u>be a representation of</u> R <u>by</u> nxn <u>matrices with elements in the sfield</u> $\mathfrak{k}$, $\alpha \to A_\alpha$. <u>Let</u> (X_k) <u>be a basis of the</u> $\mathfrak{k}$<u>-right space</u> V_n <u>of column vectors of dimension</u> n. <u>Then a representation space</u> V <u>of</u> A <u>is determined by the relations</u> $\alpha(X_k)(c_i) = (X_k)A_\alpha(c_i)$. <u>If</u> (X'_k) <u>be another basis of</u> V_n, <u>then the representation space</u> V' <u>determined by</u> $\alpha(X'_k)(d_i) = (X'_k)A_\alpha(d_i)$ <u>is</u> (R, $\mathfrak{k}$)-<u>isomorphic to</u> V.

<u>Proof</u>. For the first statement all we need to check is the associativity condition. This follows easily, for if $X \in V$, $X = (X_k)(c_i)$, $\alpha \in R$, $a \in \mathfrak{k}$,

$$\alpha(Xa) = \alpha((X_k)(c_i)a) = \alpha((X_k)(c_i a)) = (X_k)A_\alpha(c_i a),$$

while

$$(\alpha X)a = (\alpha(X_k)(c_i))a = ((X_k)A_\alpha(c_i))a = (X_k)A_\alpha(c_i a).$$

If, now, V' is determined by a second basis (X'_k), we obtain an isomorphic mapping of V' upon V by the relations $X'_k \to X_k$, $X'_k \cdot a \to X_k \cdot a$, $(X'_k)(c_i) \to (X_k)(c_i)$. From the relations for $\alpha(X_k)$, $\alpha(X'_k)$ it follows that this is an R-mapping as well as a $\mathfrak{k}$-mapping.

If, then, we do not distinguish between isomorphic representation spaces, Theorem 3.2B allows us to refer to <u>the</u> representation space corresponding to any given representation of R.

If R is an algebra over a field $\mathfrak{k}$, and (ε_k) is a basis for R, then a <u>regular</u> representation of R is obtained by

$$(2.8) \qquad \alpha(\varepsilon_k) = (\varepsilon_k)S_\alpha, \quad \alpha \to S_\alpha.$$

If R possesses a unit element, then it may be shown that $\alpha \to S_\alpha$ is an isomorphic mapping, and the matrices S_α are said to give a <u>faithful</u> representation of the algebra R.

By considering $\mathfrak{k}$-left spaces and R-right spaces and replacing column vectors by row vectors, one could, of course, obtain a set of statements in 1-1 correspondence with those we have just discussed. In particular, another regular representation is obtained for an algebra R with $\mathfrak{k}$-basis (ε_i) (here written as a column) by the relation

$$(\varepsilon_i)\alpha = T_\alpha(\varepsilon_i). \tag{2.9}$$

3. REDUCIBLE AND DECOMPOSABLE REPRESENTATION SPACES

We add here some concepts which we shall not immediately use but which should be available for interpreting the results of Chapter IX. We use the notations of the previous section.

When we refer to a subspace of a representation space we shall mean a subspace which is again an R-left and $\mathfrak{k}$-right space. A representation space V is said to be <u>reducible</u> if V has a proper subspace V_1; if V has no proper subspace, then it is called <u>irreducible</u>. A representation is called <u>reducible</u> or <u>irreducible</u> according as its corresponding representation space is reducible or irreducible. If $V \supset V_1$, V_1 an R-left and $\mathfrak{k}$-right space, then if we choose the coördinates of V by first taking a basis $X_{r+1}, \cdots, X_n$ for V_1, and then sufficient additional vectors $X_1, \cdots, X_r$ to give a basis for the whole of V, we have

$$\alpha(X_1, \cdots, X_r, X_{r+1}, \cdots, X_n) = (X_1, \cdots, X_r, X_{r+1}, \cdots, X_n) \begin{Vmatrix} A_{11}(\alpha) & \\ A_{21}(\alpha) & A_{22}(\alpha) \end{Vmatrix},$$

where $A_{11}(\alpha)$ is an $r \times r$ matrix, $A_{22}(\alpha)$ an $(n - r) \times (n - r)$ matrix. The 0-matrix in the upper right-hand corner expresses that V_1 is a subspace of V, so that for $k > r$, $\alpha X_k \in V_1$, that is, αX_k is of form $\Sigma_{\nu=r+1}^{n} X_\nu a_{\nu k}$. Here the matrices $A_{22}(\alpha)$ give a representation corresponding to the space V_1, and $A_{11}(\alpha)$ give a representation corresponding to the factor space V/V_1.

If V possesses a composition series

$$V \supset V_1 \supset V_2 \supset \cdots \supset V_{t-1} \supset (0),$$

then by adapting the coördinate system (X_k) of V to the series we obtain a representation with matrices of form

$$\begin{Vmatrix} A_{11}(\alpha) & & & \\ A_{21}(\alpha) & A_{22}(\alpha) & & \\ \vdots & & \ddots & \\ A_{t1}(\alpha) & \cdot & \cdots & A_{tt}(\alpha) \end{Vmatrix}$$

where the matrices $A_{ii}(\alpha)$ give an irreducible representation A_i corresponding to the factor space V_{i-1}/V_i. These representations A_i are called the irreducible constituents of the representation given by V. It follows from the Jordan-Hölder theorem that the irreducible constituents are uniquely determined apart from their arrangement.

A representation space V is said to be decomposable if V may be written as the direct sum of subspaces. If $V = V_1 + V_2$, then adapting the coördinate system (X_k) to this decomposition by taking for (X_k) a coördinate system $(X_1, \cdots, X_r)$ of V_1 together with a coördinate system $(X_{r+1}, \cdots, X_n)$ of V_2, we obtain the matrices of the representation determined by V in the form

$$\begin{Vmatrix} A_1(\alpha) & \\ & A_2(\alpha) \end{Vmatrix}.$$

A representation is called decomposable (indecomposable) if the corresponding space is decomposable (indecomposable).

CHAPTER IV

SEMISIMPLE RINGS

1. THE MAIN THEOREM

A ring R is said to be semisimple if it has radical zero and satisfies the minimum condition on left ideals. A ring R is said to be simple if it satisfies the minimum condition on left ideals and has no two-sided ideals other than itself and the zero ideal.

Let R be simple. Then its radical N is a two-sided ideal, and so either $N = 0$ or $N = R$. Suppose the second alternative holds (and, of course, that $R \neq 0$). N^2 is a two-sided ideal of R which cannot be all of R, else R would not be nilpotent. Hence $N^2 = R^2 = 0$. Now let $\alpha \neq 0$ be any element in R. Since all products are zero the elements $\cdots, -\alpha, 0, \alpha, 2\alpha, \cdots$ constitute a nonzero two-sided ideal in R and, therefore, are all of R. Unless the number of elements in R is a prime, the group generated by α will contain a nonzero proper subgroup, which is, of course, also a two-sided ideal. This being impossible, there remains the case that R is a cyclic group of prime order with respect to addition and $R^2 = 0$. We add to the definition of simple ring a qualification excluding this trivial case. With this convention we see that a simple ring is semisimple.

A minimal two-sided ideal of a semisimple ring can be shown to be a simple ring, and so we may call it a simple ideal. The main theorem of this chapter justifies the term semisimple by showing the structure of a semisimple ring relative to its simple ideals.

Main Theorem 4.1A. A semisimple ring R contains only a finite number of simple ideals, and is the direct sum of them. Furthermore, any two-sided ideal of R is the direct sum of those simple ideals which it contains.

The proof of this theorem is in §5 below.

2. GENERATING IDEMPOTENTS FOR LEFT IDEALS

Several of the topics discussed below are covered by the theorems of Chapter II. It is the authors' view, however, that mathematics is more appreciated and more beautiful if "towers" of theorems are avoided, and, instead, each subject is developed from its own foundation ideas. We therefore give here for several theorems the simpler proofs that are possible in the case of the semisimple ring.

Theorem 4.2A. Any left ideal $\mathfrak{l}$ *in a semisimple ring* R *contains an idempotent generator* e, *that is,* $\mathfrak{l} = Re$.

Proof. We first show that *any nonzero left ideal* $\mathfrak{l}$ *in* R *contains an idempotent*. By virtue of the minimum condition in R it is sufficient to show that any minimal left ideal $\mathfrak{l}$ contains an idempotent. For any λ in $\mathfrak{l}$, $\mathfrak{l}\lambda$ is a left ideal contained in $\mathfrak{l}$. By the minimal nature of $\mathfrak{l}$ either $\mathfrak{l}\lambda = 0$ or $\mathfrak{l}\lambda = \mathfrak{l}$. If $\mathfrak{l}\lambda = 0$ for every λ in $\mathfrak{l}$, then $\mathfrak{l}$ would be nilpotent, contrary to the semisimplicity of R. Hence for some λ we have $\mathfrak{l}\lambda = \mathfrak{l}$. Then $\mathfrak{l}$ contains an element e such that $e\lambda = \lambda$, and so $(e^2 - e)\lambda = 0$. The set of elements in $\mathfrak{l}$ which annihilate λ is a left ideal of R. It is not all of $\mathfrak{l}$, since $e\lambda = \lambda$. Hence by the minimal nature of $\mathfrak{l}$ only 0 (in $\mathfrak{l}$) annihilates λ. This proves that $e^2 = e$ or that $\mathfrak{l}$ contains the idempotent e.

The second step is to show that *in any nonzero left ideal* $\mathfrak{l}$ *there is an idempotent* e *which is annihilated (under left multiplication) only by* 0 (in $\mathfrak{l}$); that is, $\lambda e = 0$, $\lambda \varepsilon \mathfrak{l}$ implies $\lambda = 0$. With each idempotent e in $\mathfrak{l}$ we associate the left ideal $\mathfrak{m}_e$ consisting of all left annihilators of e in $\mathfrak{l}$. Let e be an idempotent for which $\mathfrak{m}_e$ is minimal. (Such an e exists because of the minimum condition on left ideals in R.) Suppose that $\mathfrak{m}_e \neq 0$; then, by the first part of the proof, $\mathfrak{m}_e$ contains an idempotent e_1. We have (from $\mathfrak{m}_e e = 0$) that $e_1 e = 0$. Set $e' = e - ee_1 + e_1$. Then $e'e = e = ee'$, $e_1 e' = e_1 = e'e_1$, and so $e'e' = e'(e - ee_1 + e_1) = e'$. Since $e'e = e$, $\mathfrak{m}_e \supseteq \mathfrak{m}_{e'}$. But, since $e_1 e = 0$ and $e_1 e' \neq 0$, $\mathfrak{m}_{e'}$ is actually a proper part of $\mathfrak{m}_e$. This contradicts the minimal nature of $\mathfrak{m}_e$. This contradiction arose from the assumption that $\mathfrak{m}_e \neq 0$. Hence $\mathfrak{m}_e = 0$, or e is annihilated from the left by no nonzero element of $\mathfrak{l}$.

The third and final step in the proof of our theorem is to show that the e (just chosen in the second step) *is a right unit for* $\mathfrak{l}$, *and* $\mathfrak{l} = Re$. Let $\lambda \varepsilon \mathfrak{l}$. Then $(\lambda - \lambda e)e = 0$. Hence $\lambda - \lambda e \varepsilon \mathfrak{m}_e$, or $\lambda = \lambda e$. This shows that e is a right unit for $\mathfrak{l}$ and that $\mathfrak{l} = \mathfrak{l}e$. Now $Re \supseteq \mathfrak{l}e = \mathfrak{l}$ since $R \supseteq \mathfrak{l}$. On the other hand, $Re \subseteq \mathfrak{l}$ since e is an element of the left ideal $\mathfrak{l}$. Hence $\mathfrak{l} = Re$.

3. UNITY ELEMENTS IN TWO-SIDED IDEALS

Theorem 4.3A. *Let* $\mathfrak{a} = Re$ *be any two-sided ideal in a semisimple ring* R. *Then* $\mathfrak{a} = eR$ *or, in other words, any generating idempotent for a two-sided ideal is a two-sided unity element for the ideal. Furthermore,* e *is uniquely determined by* $\mathfrak{a}$.

Proof. Since $\mathfrak{a}$ is a right ideal, the set of all ξ in $\mathfrak{a}$ for which $e\xi = 0$ is a right ideal $\mathfrak{r}$ of R. Now apply $e\mathfrak{r} = 0$ and $\mathfrak{r}e = \mathfrak{r}$ to the product $\mathfrak{r}\mathfrak{r}$, giving $\mathfrak{r}\mathfrak{r} = (\mathfrak{r}e)\mathfrak{r}$ $= \mathfrak{r}(e\mathfrak{r}) = 0$, that is, $\mathfrak{r}$ is a nilpotent right ideal. $\mathfrak{r} + R\mathfrak{r}$ is a two-sided nilpotent ideal, and so must be zero, else R is not semisimple. But $\mathfrak{r} + R\mathfrak{r} = 0$ if and only if $\mathfrak{r} = 0$. (Or one can argue directly that $\mathfrak{r} = 0$, since a ring without radical, and with minimum condition on left ideals, can have neither left nor right nilpotent left ideals other than 0.) Clearly $e(\alpha - e\alpha) = 0$ for α in $\mathfrak{a}$, and so $\alpha - e\alpha \in \mathfrak{r}$, that is, $\alpha - e\alpha = 0$. This shows that e is a two-sided unity element for $\mathfrak{a}$. To establish the uniqueness of e we suppose that e' is also a unity element for $\mathfrak{a}$ and "set up a competition" between e and e'. We have $e' = e'e = e$. Indeed, we note that, if any ring has a left unity element e' and a right unity element e, they must coincide.

The case $\mathfrak{a} = R$ of the present theorem is the important

Corollary 4.3B. *Any semisimple ring has a (two-sided) unity element (denoted by 1).*

4. THE CENTER OF A SEMISIMPLE RING

The set of elements ξ in a ring R which are commutative with all the elements of R is called the *center*, C, of R. It is easy to see that the center of a ring is a commutative subring.

Theorem 4.4A. *Let* R *be a semisimple ring with center* C. *Then any idempotent* e *in* C *is unity element for a two-sided ideal* $\mathfrak{a} = Re$ *of* R; *and, conversely, the unity element* e *of any two-sided ideal* $\mathfrak{a}$ *of* R *is in* C.

Proof. In other words, the theorem states that there is a 1-1 correspondence between two-sided ideals $\mathfrak{a}$ of R and idempotents e in C, this correspondence being given by $e \rightarrow \mathfrak{a} = Re$. It is clear that, if $e\alpha = \alpha e$ for all α in R, then $\mathfrak{a} = Re$ $(= eR)$ is a two-sided ideal of R. Indeed, this would still be true with e replaced by any element of C. Conversely, if e is unity element for a two-sided ideal $\mathfrak{d}$, then, for all α in R, $e\alpha$ and αe lie in $\mathfrak{d}$. Hence $e\alpha = (e\alpha)e = e(\alpha e) = \alpha e$, that is, $e \in C$.

Theorem 4.4B. *Let $\mathfrak{d} = Re$ be a two-sided ideal in a semisimple ring R. Then R is the direct sum of $\mathfrak{d}$ and a second, uniquely determined, two-sided ideal $\mathfrak{d}'$.*

Proof. We set $\mathfrak{d}' = R(1 - e)$. Any element α in R can be written $\alpha = \alpha e + \alpha(1 - e)$, which shows that $R = \mathfrak{d} + \mathfrak{d}'$. To see that the sum is direct we suppose that $0 = \alpha + \alpha'$, where $\alpha \in \mathfrak{d}$ and $\alpha' \in \mathfrak{d}'$. Then multiplying by e we get, since $e\mathfrak{d}' = e(1 - e)R = 0$ and $e\alpha = \alpha$, that $\alpha = 0$ and therefore also $\alpha' = 0$.

Lemma 4.4C. *Any two-sided ideal $\mathfrak{d}$ of a semisimple ring R is itself a semisimple ring.*

Proof. The defining conditions for semisimplicity of $\mathfrak{d}$ will follow from those in R if we can show that every left ideal $\mathfrak{l}$ in $\mathfrak{d}$ is also a left ideal in R. Let $\mathfrak{d} = eR$; since $\mathfrak{l} \subseteq \mathfrak{d}$, we have $\mathfrak{l} = e\mathfrak{l}$. Therefore, $R\mathfrak{l} = (Re)\mathfrak{l} = \mathfrak{d}\mathfrak{l} \subseteq \mathfrak{l}$, and so $\mathfrak{l}$ is a left ideal in R.

5. PROOF OF THE MAIN THEOREM

Lemma 4.5A. *Let $\mathfrak{d}_1, \cdots, \mathfrak{d}_n$ be distinct simple ideals of a semisimple ring R. Then the sum $\mathfrak{d} = \mathfrak{d}_1 + \cdots + \mathfrak{d}_n$ is direct.*

Proof. Let e_i be the unity element of $\mathfrak{d}_i$. The intersection of two simple ideals is a two-sided ideal, and so must be zero. Hence $e_i\mathfrak{d}_j = 0$ if $i \neq j$. Now suppose that $0 = \alpha_1 + \cdots + \alpha_n$, where $\alpha_i \in \mathfrak{d}_i$. Then multiplication by e_i gives $0 = e_i\alpha_i = \alpha_i$. This shows that 0 has only one expression and, therefore, that the sum is direct, as claimed.

Now, to prove the main theorem (4.1A), we consider the set of ideals $R(1 - e)$, where e runs over the set of all finite sums of unity elements of simple ideals. Because of the minimum condition in R we can select $e = e_1 + \cdots + e_n$, so that $R(1 - e)$ is minimal. Then $R = Re_1 + \cdots + Re_n + R(1 - e)$ is a direct decomposition of R. If $R(1 - e) \neq 0$, then it contains a simple ideal Re_{n+1}. Set $e' = e + e_{n+1}$, and then the direct splitting $R(1 - e) = R(1 - e') + Re_{n+1}$ (true by Theorem 4.4B) shows that $R(1 - e')$ is properly contained in $R(1 - e)$, contrary to the minimal nature of $R(1 - e)$. This contradiction comes from the assumption that $R(1 - e)$ was not zero.

Hence R has only a finite number of simple ideals and is the direct sum of them. The last part of the main theorem now follows from the first part and Lemma 4.4C.

This completes the proof of the main theorem, but before turning to the theory of simple rings we show in just a bit more detail exactly how a semisimple ring is determined by its simple ideals. We have $R = \mathfrak{a}_1 + \cdots + \mathfrak{a}_n$, where each $\mathfrak{a}_i$ is a simple ring and the sum is direct. Let $\alpha = \alpha_1 + \cdots + \alpha_n$ and $\alpha' = \alpha_1' + \cdots + \alpha_n'$, where α_i, $\alpha_i' \in \mathfrak{a}_i$. Then $\alpha + \alpha' = (\alpha_1 + \alpha_1') + \cdots + (\alpha_n + \alpha_n')$ and $\alpha\alpha' = \alpha_1\alpha_1' + \cdots + \alpha_n\alpha_n'$. In other words, to make computations in R, we need know only how to make computations in its simple ideals. One could indicate this by considering $\alpha_1, \cdots, \alpha_n$ as components of coördinates for α and writing $\alpha = (\alpha_1, \cdots, \alpha_n)$. Then addition and multiplication of two elements of R are carried out by adding and multiplying corresponding components.

CHAPTER V

SIMPLE RINGS

1. WEDDERBURN'S THEOREM

The main theorem on the structure of simple rings is due to Wedderburn, and reduces the study of simple rings to that of sfields.

Wedderburn's Theorem 5.1A. _Any simple ring_ R _is isomorphic to the ring of all_ m-_rowed square_ $\mathfrak{k}$-_matrices, where the sfield_ $\mathfrak{k}$ _and the integer_ m _are uniquely defined by_ R. _Conversely, for any integer_ m _and sfield_ $\mathfrak{k}$ _the set of all_ m-_rowed square_ $\mathfrak{k}$-_matrices is a simple ring_.

The present chapter has as its chief content the proof of Wedderburn's theorem. A number of other important properties of simple rings are discussed. In several instances it is found convenient to state and prove results directly for semisimple rings, rather than to follow strictly our general program of deducing results for semisimple rings from those of simple rings. A justification for this deviation is that frequently in applications (for instance group rings) it is with a semisimple ring rather than with its simple ideals that we are primarily concerned.

2. THE CONVERSE PART OF WEDDERBURN'S THEOREM

Let $\mathfrak{k}$ be a sfield, It satisfies the minimum condition since it has only itself and 0 as ideals. Hence, by Theorem 2.2E, the ring R of all m-rowed square $\mathfrak{k}$-matrices satisfies the minimum condition.

We denote by e_{ik} the matrix with 1 (unity of $\mathfrak{k}$) at the intersection of the i-th row and k-th column and 0 elsewhere. Then any element α in R can be written uniquely in the form $\alpha = \Sigma a_{\nu\mu} e_{\nu\mu}$, $a_{ik} \varepsilon \mathfrak{k}$. Suppose $\mathfrak{a}$ is any nonzero two-sided ideal of R, and let $\alpha \neq 0$ be an element of $\mathfrak{a}$. Since $\alpha \neq 0$, some coefficient, say a_{ik}, is not zero. Since $\mathfrak{a}$ is two-sided, it will contain $(b_{h\lambda} a_{ik}^{-1} e_{hi}) \alpha e_{k\lambda} = b_{h\lambda} e_{h\lambda}$ for any h and λ and element $b_{h\lambda}$ in k. Hence $\mathfrak{a} = R$; and so R is a simple ring.

3. DECOMPOSITION OF A SEMISIMPLE RING INTO LEFT IDEALS

Theorem 5.3A. Let $\mathfrak{l}$ be a left ideal in a semisimple ring R, and suppose that $\mathfrak{l}_1$ is a subideal of $\mathfrak{l}$. Then R contains an ideal $\mathfrak{l}_2$ such that $\mathfrak{l} = \mathfrak{l}_1 + \mathfrak{l}_2$ (direct sum).

In the preceding chapter we proved the analogous theorem for two-sided ideals and, moreover, were able to establish uniqueness of the summands. For the decomposition into left ideals uniqueness no longer exists. Indeed, the development below will show that there are as many choices for $\mathfrak{l}_2$ as there are generating idempotents for $\mathfrak{l}_1$.

Proof. Let $\mathfrak{l} = Re$ and $\mathfrak{l}_1 = Re_1$, where e and e_1 are idempotents (cf. Theorem 4.2A). Write any element λ of $\mathfrak{l}$ in the form $\lambda = \lambda_1 + \lambda_2$, where $\lambda_1 = \lambda e_1 \varepsilon \mathfrak{l}_1$ and $\lambda_2 = \lambda - \lambda e_1$ is a left annihilator of e_1. The set of elements in $\mathfrak{l}$ which are left annihilators of e_1 is a left ideal $\mathfrak{l}_2$ of R. The above splitting of λ shows that $\mathfrak{l} = \mathfrak{l}_1 + \mathfrak{l}_2$. The equation $\lambda e_1 = \lambda_1$ shows that the splitting of λ is unique, or, in other words, that the sum is direct.

This proves the theorem, but it is of some interest to obtain generating idempotents for $\mathfrak{l}_1$ and $\mathfrak{l}_2$, which are orthogonal to each other and whose sum is the given idempotent generator e of $\mathfrak{l}$. We have $e = e_1' + e_2'$, where $e_1' = ee_1 \;\varepsilon\; \mathfrak{l}_1$ and $e_2' = e - ee_1 \;\varepsilon\; \mathfrak{l}_2$. Multiplication by e_1' (on the left) gives $e_1'e = e_1' = e_1'e_1' + e_1'e_2'$. Since $\mathfrak{l}$ is the direct sum of $\mathfrak{l}_1$ and $\mathfrak{l}_2$, this requires $e_1' = e_1'e_1'$ and $e_1'e_2' = 0$; similarly, $e_2' = e_2'e_2'$ and $e_2'e_1' = 0$. The orthogonality of e_1' and e_2' shows that they commute with their sum e. That e_1' is a generating idempotent for $\mathfrak{l}_1$ follows from the equations $e_1' = ee_1$ and $e_1 = e_1e_1 = (e_1e)e_1 = e_1(ee_1) = e_1e_1'$. Similarly for e_2' and $\mathfrak{l}_2$.

We say that an idempotent is primitive if it cannot be written as a sum of two nonzero orthogonal idempotents. The above discussion shows that in a semisimple ring a left ideal $\mathfrak{l}$ is minimal if and only if it has a primitive generating idempotent.

Corollary 5.3B. Any semisimple ring R can be written as the direct sum of a finite number of minimal left ideals.

Proof. By the minimum condition, R contains some minimal left ideal $\mathfrak{l}_1$. Apply the theorem 5.3A to get $R = \mathfrak{l}_1 + \mathfrak{l}_1'$(direct sum), where $\mathfrak{l}_1'$ is a left ideal. Again by the minimum condition, $\mathfrak{l}_1'$ is either 0 or contains a minimal left ideal $\mathfrak{l}_2$, and then we get $R = \mathfrak{l}_1 + \mathfrak{l}_2 + \mathfrak{l}_2'$ (direct sum). This process must stop with $R = \mathfrak{l}_1 + \cdots + \mathfrak{l}_n$ (direct sum), for some n because the set R, $\mathfrak{l}_1'$, $\mathfrak{l}_2'$, $\cdots$ is a descending chain of left ideals in R. (To avoid the use of the choice axiom here we may consider the set of all direct sums $\mathfrak{l}$ of a finite number of minimal left ideals. For each such $\mathfrak{l}$ we find a complement $\mathfrak{l}'$ such that $R = \mathfrak{l} + \mathfrak{l}'$ (direct sum). There is a minimal ideal in the set of all $\mathfrak{l}'$. By the preceding argument this minimal $\mathfrak{l}'$ must be zero, and so we are done.)

If we consider R as an R-space, then the minimal left ideals are irreducible R-spaces. We have just obtained one decomposition of the space R into a direct sum of irreducible R-spaces. By Theorem 1.6C the components in this decomposition are unique, to within isomorphisms. In other words, any minimal left ideal in R is isomorphic to one of $\mathfrak{l}_1, \cdots, \mathfrak{l}_n$.

Lemma 5.3C. *Let $\mathfrak{l}$ be a left ideal in any ring R. Then the sum $\mathfrak{s}$ of all left ideals homomorphic to $\mathfrak{l}$ (as R-spaces) is a two-sided ideal of R.*

Proof. Let σ: $\lambda \to \lambda' = \sigma(\lambda)$ be an R-homomorphism of $\mathfrak{l}$ onto $\mathfrak{l}'$ (that is, σ is an additive mapping of $\mathfrak{l}$ onto $\mathfrak{l}'$, with the further property $\sigma(\xi\lambda) = \xi\sigma(\lambda)$ for any $\xi \varepsilon R$ and $\lambda \varepsilon \mathfrak{l}$). Let $\alpha \varepsilon R$. Then $\mathfrak{l}'\alpha$ is a left ideal of R, R-homomorphic to $\mathfrak{l}$ under the mapping σ': $\lambda \to \lambda'\alpha$; that is, $\sigma'(\lambda) = \sigma(\lambda)\alpha$. This shows that the set of R-homomorphic images of $\mathfrak{l}$ is closed under right multiplication by elements of R, or, in other words, that the sum $\mathfrak{s}$ is a right ideal. But $\mathfrak{s}$ is a sum of left ideals and so is two-sided.

Theorem 5.3D. *Let R be a simple ring. Then all its minimal left ideals are isomorphic, as R-spaces.*

Proof. Since a simple ring is semisimple, we can write $R = \mathfrak{l}_1 + \cdots + \mathfrak{l}_n$, where each $\mathfrak{l}_i$ is a minimal left ideal and the sum is direct. Suppose the terms in this sum so arranged that $\mathfrak{l}_1, \cdots, \mathfrak{l}_m$ are R-isomorphic to $\mathfrak{l}_1$, and no $\mathfrak{l}_i$ with $i > m$ is R-isomorphic to $\mathfrak{l}_1$. By Theorem 1.6C (with V = R) any left ideal R-isomorphic to $\mathfrak{l}_1$ is contained in $\mathfrak{s} = \mathfrak{l}_1 + \cdots + \mathfrak{l}_m$. Since $\mathfrak{l}_1$ is minimal, its nonzero homomorphic images are actually isomorphic images. Hence $\mathfrak{s}$ is the sum of all left ideals R-homomorphic to $\mathfrak{l}_1$. By the preceding lemma $\mathfrak{s}$ is a two-sided ideal (obviously not zero) of R. Since R is simple, we must have $\mathfrak{s} = R$.

But then $m = n$, and our theorem is proved.

Theorem 5.3E. *Let V be an irreducible space over a simple ring R. Then, if $RV \neq 0$, V is isomorphic to any minimal left ideal $\mathfrak{l}$ of R.*

Proof. If $\mathfrak{l}V = 0$, then the annihilator $\mathfrak{a}$ of V would be a two-sided ideal of R containing $\mathfrak{l}$; therefore not the zero ideal, and hence all of R. But part of our hypothesis was $RV \neq 0$. Hence V contains a vector X not annihilated by $\mathfrak{l}$. Since $\mathfrak{l}$ is a left ideal, $R(\mathfrak{l}X) \subseteq \mathfrak{l}X$, and so $\mathfrak{l}X$ is a nonzero R-space contained in V. Hence $\mathfrak{l}X = V$; and so the mapping $\lambda \rightarrow \lambda X$ is an R-homomorphism of $\mathfrak{l}$ onto V. It is, moreover, an isomorphism, since otherwise the elements of $\mathfrak{l}$ which were mapped into 0 in V would be a proper, nonzero, subideal of $\mathfrak{l}$, contrary to the minimal nature of $\mathfrak{l}$.

Note that Theorem 5.3D is a direct corollary to Theorem 5.3E. However, the proof of the former theorem shows how to compute the simple ideals of any semisimple ring. Indeed, *if $R = \mathfrak{l}_1 + \cdots + \mathfrak{l}_n$ is a (direct) decomposition of the semisimple ring R into minimal left ideals, then every simple ideal of R can be obtained by adding all left ideals in the set $\mathfrak{l}_1, \cdots, \mathfrak{l}_n$ that are isomorphic (as R-spaces) to some given one, $\mathfrak{l}_i$.*

Theorem 5.3F. *Let V be a vector space over a semisimple ring R and suppose that $RV = V$. Then V is the sum of its irreducible R-spaces.*

Proof. All that we have to show is that any vector of V is contained in a finite sum of irreducible R-spaces. Suppose $R = \mathfrak{l}_1 + \cdots + \mathfrak{l}_n$ is a direct decomposition of R into minimal left ideals, and let X be any vector in V. The equation $RV = V$ carries with it the fact that the unity element of R is the identity operator on V. Hence $X = 1X \in RX = \mathfrak{l}_1X + \cdots + \mathfrak{l}_nX$, where each $\mathfrak{l}_iX$ is either an irreducible R-space or zero.

Lemma 5.3G. *Let R be any ring, and let V be an R-space with minimum condition. Then any sequence of subspaces of V whose sum is direct must be of finite length.*

Proof. This is practically a restatement of the minimum condition. Suppose that $V_1, V_2, \cdots$ is any infinite sequence of subspaces of V whose sum is direct. Set $V_i' = V_i + V_{i+1} + \cdots$; V_i' is not contained in V_{i+1}', else the sum of the V's would not be direct. Hence V_{i+1}' is a proper part of V_i', for $i = 1, 2, \cdots$. This contradicts the hypothesis that the minimum condition holds in V; hence every sequence of subspaces whose sum is direct has finite length.

Theorem 5.3H. Let R *be a semisimple ring, and let* V *be an* R*-space with minimum condition and with* $RV = V$. *Then* V *is the direct sum of a finite number of irreducible* R*-spaces.*

Proof (using choice axiom). Let V_1 be an irreducible subspace of V. If $V_1 \neq V$, then, by Theorem 5.3F, V contains a second irreducible subspace V_2 distinct from V_1. $V_1 + V_2$ is a direct sum (for if $V_1 \cap V_2 \neq 0$ one of V_1, V_2 is not irreducible; cf. Theorem 1.4C), if $V_1 + V_2 \neq V$, we find an irreducible V_3 such that $V_3 \cap (V_1 + V_2) = 0$; so that the sum $V_1 + V_2 + V_3$ is direct. Proceeding thus we must get $V = V_1 + \cdots + V_n$, (direct sum), since otherwise we would arrive at a contradiction of the preceding lemma.

4. MINIMAL LEFT IDEALS

Theorem 5.4A. Let R *be a ring without nilpotent ideals. Then* $\mathfrak{l} = Re$ *is a minimal left ideal if and only if* eRe *is a sfield.*

Proof:

Only if. For any idempotent e of R the set eRe is a ring with unity element e. Thus to show that eRe is a sfield when $\mathfrak{l}$ is minimal we need only find an inverse (under multiplication) for each nonzero element. Suppose $a \neq 0$ is an element of eRe. Then $0 \neq Ra \subseteq \mathfrak{l}$, since $a \in \mathfrak{l}$. But $\mathfrak{l}$ is a minimal left ideal, hence $Ra = \mathfrak{l} = Re$. Therefore $eRa = eRe$; since $a = eae = ea$, this gives $(eRe)a = (eRe)$, and so a has at least one left inverse a^{-1} such that $a^{-1}a = e$; this shows that eRe is a sfield.

If. Let eRe be a sfield, set $\mathfrak{l} = Re$, and suppose that $\mathfrak{l}'$ is a nonzero left ideal contained in $\mathfrak{l}$. If $e\lambda' = 0$ for every λ' in $\mathfrak{l}'$, then $\mathfrak{l}\mathfrak{l}' = \mathfrak{l}'\mathfrak{l}' = 0$, contrary to the hypothesis that R had no nilpotent ideals. Hence $\mathfrak{l}'$ contains a nonzero element of the form $e\lambda' = e\lambda'e$. Now, since eRe is a sfield, $e\lambda'e$ has an inverse $e\lambda_1e$. Hence $R(e\lambda'e)$ contains $(e\lambda_1e)(e\lambda'e) = e$; but, since $e\lambda'e \in \mathfrak{l}'$, this gives $e \in R(e\lambda'e) \subseteq \mathfrak{l}'$; and, finally, $\mathfrak{l} = Re \subseteq R\mathfrak{l}' \subseteq \mathfrak{l}'$, that is, $\mathfrak{l} = \mathfrak{l}'$. This shows that $\mathfrak{l}$ is minimal.

Corollary 5.4B. Let R be any ring without nilpotent ideals. If $\mathfrak{l} = Re$ *is a minimal left ideal, then* $\mathfrak{r} = eR$ *is a minimal right ideal.*

Proof. The above theorem is clearly true with "left" replaced by "right" throughout, since the hypotheses on R are all two-sided.

5. THE DIRECT PART OF WEDDERBURN'S THEOREM

Theorem 5.5A. Let $\mathfrak{l} = Re$ *be a minimal left ideal in the simple ring R. Set* $\mathfrak{k} = eRe$ *and* $\mathfrak{r} = eR$. *Suppose* $R = \mathfrak{l}_1 + \cdots + \mathfrak{l}_m$ *(direct sum), where the* $\mathfrak{l}_i$ *are minimal left ideals. Then* $\mathfrak{r}$ *has dimension m over the sfield* $\mathfrak{k}$.

Proof. That all the hypotheses of the present theorem can be realized for any simple ring R has been established in the last two sections. We use also the fact that $\mathfrak{r}$ is a minimal right ideal (cf. 5.3B, 5.3D, 5.4A, 5.4B).

We first remark that $eReeR \subseteq eR = \mathfrak{r}$, so that $\mathfrak{r}$ is a left $\mathfrak{k}$-space. Now suppose that $\alpha_1, \cdots, \alpha_n$ are elements of $\mathfrak{r}$, linearly independent with respect to $\mathfrak{k}$. Suppose that

$$(5.1) \qquad \xi_1\alpha_1 + \cdots + \xi_n\alpha_n = 0, \qquad \xi_1, \cdots, \xi_n \in R.$$

We shall show that (5.1) implies $\xi_i\alpha_i = 0$, $i = 1, \cdots, n$. Set $\lambda_i = \xi_i e$, then we have from (5.1) (since $\alpha_i = e\alpha_i$):

$$\lambda_1\alpha_1 + \cdots + \lambda_n\alpha_n = 0, \qquad \lambda_1, \cdots, \lambda_n \in Re = \mathfrak{l}.$$

For each i, $\mathfrak{l}\alpha_i$ is a minimal left ideal different from zero, since $e\alpha_i = \alpha_i \neq 0$. Now suppose $\lambda_1\alpha_1 (= \xi_1\alpha_1) \neq 0$. Then $\lambda_1\alpha_1 \in \mathfrak{l}\alpha_2 + \cdots + \mathfrak{l}\alpha_n$, or $\mathfrak{l}\alpha_1 \subseteq \mathfrak{l}\alpha_2 + \cdots + \mathfrak{l}\alpha_n$ (since a left ideal either contains a minimal left ideal or has no element $\neq 0$ in common with it). In particular, $\alpha_1 \in \mathfrak{l}\alpha_1$, and so there are elements $\lambda_2', \cdots, \lambda_n'$ in $\mathfrak{l}$ such that

$$\alpha_1 + \lambda_2'\alpha_2 + \cdots + \lambda_n'\alpha_n = 0.$$

Now multiplication by e on the left gives

$$e\alpha_1 + e\lambda_2'\alpha_2 + \cdots + e\lambda_n'\alpha_n = 0.$$

But e, $e\lambda_2' (= e\lambda_2' e), \cdots, e\lambda_n'$ all lie in $\mathfrak{k}$. Since $e \neq 0$, this contradicts the $\mathfrak{k}$-independence of the α's. This contradiction arises from the assumption that (5.1) is true without each term $\xi_i\alpha_i$ being zero. From the same argument we conclude also that $\mathfrak{l}\alpha_1 + \cdots + \mathfrak{l}\alpha_n$ is a direct sum of minimal left ideals, thus proving that $n \leq m$ (also proving that $\mathfrak{r}$ has finite $\mathfrak{k}$-dimension).

Conversely, suppose $\mathfrak{r} = \mathfrak{k}\alpha_1 + \cdots + \mathfrak{k}\alpha_n$. $\mathfrak{l}\mathfrak{r}$, as a nonzero two-sided ideal of the simple ring R must equal R. Now $\mathfrak{l}\mathfrak{k} = \mathfrak{l}$, so $R = \mathfrak{l}\mathfrak{r} = \mathfrak{l}\alpha_1 + \cdots + \mathfrak{l}\alpha_n$, or $n \geq m$. Hence $n = m$.

Recapitulating, $\mathfrak{r}$ is a left $\mathfrak{k}$-space of dimension m, with $\mathfrak{k}$-basis $\alpha_1, \cdots, \alpha_m$. We now show that R is isomorphic to the ring $M_m(\mathfrak{k})$ of all m-rowed square $\mathfrak{k}$-matrices by considering the effect on $\mathfrak{r}$ of multiplication by R on the right. Let $\xi \in R$ and suppose that $\alpha_i\xi = \beta_i = \Sigma a_{i\nu}\alpha_\nu$. Then from the equations

$$\begin{Vmatrix} \alpha_1 \\ \cdot \\ \cdot \\ \cdot \\ \alpha_m \end{Vmatrix} \xi = \begin{Vmatrix} \beta_1 \\ \cdot \\ \cdot \\ \cdot \\ \beta_m \end{Vmatrix} = A_\xi \begin{Vmatrix} \alpha_1 \\ \cdot \\ \cdot \\ \cdot \\ \alpha_m \end{Vmatrix}, \qquad \text{where} \quad A_\xi = \begin{Vmatrix} a_{11} & \cdots & a_{1m} \\ & \cdots\cdots & \\ a_{m1} & \cdots & a_{mm} \end{Vmatrix}$$

we see that the mapping $\xi \to A_\xi$ is a homomorphism of R into the ring $M_m(\mathfrak{k})$ of $\mathfrak{k}$-mappings of $\mathfrak{r}$ into itself. Now to complete the proof of Wedderburn's theorem (aside from the uniqueness of $\mathfrak{k}$ and m, which we consider later), we need only show that this mapping is an isomorphism onto $M_m(\mathfrak{k})$. This is accomplished by the following theorem.

Theorem 5.5B. Let $\alpha_1, \cdots, \alpha_m$ be a $\mathfrak{k}$-basis for the minimal right idea $\mathfrak{r} = eR$ of the simple ring R, and let $\beta_1, \cdots, \beta_m$ be any elements of $\mathfrak{r}$. Then there is one and only one element ξ in R such that $\beta_i = \alpha_i\xi$, $i = 1, \cdots, m$.

Proof:

Only one. The set of ξ for which $\alpha_i\xi = 0$, $i = 1, \cdots, m$, that is, for which $\mathfrak{r}\xi = 0$, is a two-sided ideal $\mathfrak{o}$ of R. Since R contains a unity element, $\mathfrak{o} \neq R$; and so $\mathfrak{o} = 0$. Now suppose $\alpha_i\xi = \alpha_i\xi'$, $i = 1, \cdots, m$. Then $\xi - \xi' \in \mathfrak{o}$, that is, $\xi = \xi'$.

At least one. We show that for each i from 1 to m there is an element ξ_i such that $\alpha_i\xi_i = \beta_i$ and $\alpha_j\xi_i = 0$ if $i \neq j$; and then set $\xi = \xi_1 + \cdots + \xi_m$. It is sufficient to show how, say, ξ_m is determined. Let $\mathfrak{l}' = \mathfrak{l}\alpha_1 + \cdots + \mathfrak{l}\alpha_{m-1}$, and let e' be a generating idempotent for $\mathfrak{l}'$. Then $1 - e'$ is a right annihilator of $\mathfrak{l}'$ (since $e'(1 - e') = 0$). Hence $\alpha_1(1 - e') = \cdots = \alpha_{m-1}(1 - e') = 0$. We cannot have $\alpha_m(1 - e') = 0$, for then $R(1 - e')$ would be zero, whereas $1(1 - e') = 1 - e' \neq 0$ since $\mathfrak{l}'$ is not all of R. Hence $\alpha_m(1 - e')R$ is a nonzero right ideal contained in $\mathfrak{r}$. Since $\mathfrak{r}$ is minimal this implies that $\alpha_m(1 - e')R = \mathfrak{r}$; in particular, R contains an element η_m such that $\alpha_m(1 - e')\eta_m = \beta_m$.

Now $\xi_m = (1 - e')\eta_m$ is the element we are looking for.

6. THE UNIQUENESS PART OF WEDDERBURN'S THEOREM

Before embarking on the uniqueness proofs we introduce the concept of inverse (or reciprocal) ring. Let R be any ring. To each element α in R we relate a new element α^* in a set R^*. We say that $\alpha^* = \beta^*$ if and only if $\alpha = \beta$. We make R^* into a ring by the definitions $\alpha^* + \beta^* = \gamma^*$ if $\alpha + \beta = \gamma$, $\alpha^*\beta^* = \gamma^*$ if $\beta\alpha = \gamma$. We say that R^* is the ring inverse (or reciprocal) to R. Any ring T isomorphic to R^* is said to be inverse isomorphic to R. Clearly $(R^*)^*$ is isomorphic to R, so that we identify it with R, thus making the *-process involutory. The same definitions apply to sfields (considered as a special case of rings).

Theorem 5.6A. *Let V be any irreducible vector space over a simple ring R, with $RV \neq 0$. Then the set of R-homomorphisms of V into itself constitutes a sfield inverse isomorphic to $\mathfrak{k} = eRe$, where e is an idempotent generator for a minimal left ideal $\mathfrak{l} = Re$ in R.*

Proof. We remark, first, that the purpose of the hypothesis $RV \neq 0$ is to exclude the case that V is a cyclic group of prime order annihilated by R, and the case $V = 0$. Since V is irreducible, and $RV \neq 0$, there is a vector X_0 in eV such that $V = \mathfrak{l}X_0$, and the mapping $\lambda \to \lambda X_0$ is an R-isomorphism between $\mathfrak{l}$ and V (cf. Theorem 5.3E). Let σ: $X \to X' = \sigma(X)$ be any R-homomorphism of V into itself. Suppose that $\sigma(X_0) = \varkappa X_0$, $\varkappa$ in $\mathfrak{l}$. Since $\mathfrak{l}$ and V are isomorphic, $\varkappa$ is uniquely defined by $\sigma(X_0)$. Then the arbitrary vector $X = \lambda X_0$ in V ($\lambda \in \mathfrak{l}$) has the image $\sigma(X) = \sigma(\lambda X_0) = \lambda\varkappa X_0$. Thus the homomorphism σ is completely defined by the image of the single vector X_0, or, indeed, by the element $\varkappa$ of $\mathfrak{l}$. This justifies the notation $\sigma_\varkappa$ for the homomorphism of V which maps X_0 into $\varkappa X_0$. We now study the range of $\varkappa$ (in $\mathfrak{l}$). Since $X_0 \in eV$, $eX_0 = X_0$, and $\varkappa X_0 = \sigma_\varkappa(X_0) = \sigma_\varkappa(eX_0) = e\varkappa X_0$, or $\varkappa = e\varkappa$. Since $\varkappa \in \mathfrak{l} = Re$, $\varkappa = \varkappa e$, and so $\varkappa (= e\varkappa e)$ is an element of $\mathfrak{k} = eRe$. This shows that each R-homomorphism of V into itself corresponds to an element of $\mathfrak{k} = eRe$. Conversely, it is clear that for any $\varkappa$ in $\mathfrak{k}$ the mapping $\sigma_\varkappa$ which sends X_0 into $\varkappa X_0$ and $X = \lambda X_0$ into $X' = \lambda\varkappa X_0$ is a homomorphism of V into itself. Furthermore, distinct elements of eRe give distinct homomorphisms (for σ_0 is the only 0-homomorphism). We define addition and multiplication of homomorphisms by linearity and iteration, respectively. Then, since $\sigma_{\varkappa}(X_0) + \sigma_{\varkappa'}(X_0) = (\varkappa + \varkappa')X_0 = \sigma_{\varkappa+\varkappa'}(X_0)$ and $\sigma_{\varkappa'}(\sigma_{\varkappa}(X_0)) = \sigma_{\varkappa'}(\varkappa X_0) = \varkappa\varkappa' X_0 = \sigma_{\varkappa\varkappa'}(X_0)$, we see that the mapping $\varkappa \to \sigma_{\varkappa}$ is an inverse isomorphism between the sfield $\mathfrak{k}$ and the set of R-homomorphisms of V into itself.

Since (Theorem 5.3E) all irreducible spaces over a simple ring are isomorphic, this proves

Corollary 5.6B. If e and e' are primitive idempotents of a simple ring R, then eRe and e'Re' are isomorphic sfields.

Now suppose that R is isomorphic to the total matrix ring $M_m(\mathfrak{k})$. Let $e(= e_{11})$ be the element in R whose image in $M_m(\mathfrak{k})$ has zeros everywhere except for the unit element of $\mathfrak{k}$ in the upper left-hand corner. Clearly eRe is isomorphic to $\mathfrak{k}$, and so, by Theorem 5.4A, e is a primitive idempotent. Now suppose that R is also isomorphic to $M_{m'}(\mathfrak{k}')$. We can, similarly, obtain an idempotent e' such that e'Re' is isomorphic to $\mathfrak{k}'$. But now, by the above corollary, $\mathfrak{k}$ is isomorphic to $\mathfrak{k}'$; and then, obviously, $m = m'$. This establishes the uniqueness of m and $\mathfrak{k}$ and so completes the proof of Wedderburn's theorem.

7. RINGS OF HOMOMORPHISMS

An important step in the above proof of Wedderburn's theorem was the study of homomorphisms of an irreducible vector space into itself. We devote the rest of this chapter to a more general study of homomorphisms.

Let R be a ring and V an R-space such that $\rho V = 0$ ($\rho \in R$) requires $\rho = 0$; in other words, R is faithfully represented by its operation on V. We shall designate R-homomorphisms of V (into itself) as right-hand operators. If $X \to X' = (X)\rho'$ is an R-homomorphism, then $(\rho X)\rho' = \rho[(X)\rho']$. This justifies us in dropping the function notation; so we write simply $X' = X\rho'$ for the mapping. If ρ'_1 and ρ'_2 are two R-homomorphisms of V, then we define the sum homomorphism $\rho'_1 + \rho'_2$ by the equation $X(\rho'_1 + \rho'_2) = X\rho'_1 + X\rho'_2$; and we define the product homomorphism $\rho'_1\rho'_2$ by iteration, that is, $X\rho'_1\rho'_2 = (X\rho'_1)\rho'_2$. The addition and multiplication thus defined satisfy both distributive laws (but for different reasons). For the left distributive law we have

$$X\rho'_3(\rho'_1 + \rho'_2) = (X\rho'_3)(\rho'_1 + \rho'_2) = X\rho'_3\rho'_1 + X\rho'_3\rho'_2 = X(\rho'_3\rho'_1 + \rho'_3\rho'_2)$$

by the definitions of multiplication and addition; and

for the other one

$$X(\rho_1' + \rho_2')\rho_3' = [X(\rho_1' + \rho_2')]\rho_3' = [X\rho_1' + X\rho_2']\rho_3'$$

by the definitions of multiplication and addition. To get rid of the brackets we use the fact that ρ_3' (as an R-homomorphism) is a linear mapping of V into itself, giving

$$[X\rho_1' + X\rho_2']\rho_3' = X\rho_1'\rho_3' + X\rho_2'\rho_3' = X(\rho_1'\rho_3' + \rho_2'\rho_3').$$

The set R' of all the homomorphisms ρ', with addition and multiplication defined as above, is therefore a ring which is right operator domain for V. Because we have written the homomorphisms as right factors of V, the ring R', thus defined, is actually inverse isomorphic to the ring of homomorphisms.

It may avoid confusion in the following discussion to go into a bit more detail here. One could denote R-homomorphisms of V by a functional notation $\sigma\colon X \to X' = \sigma(X)$. Then the equation $\sigma(\rho X) = \rho \cdot \sigma(X)$ expresses the fact that σ is an R-homomorphism of V. The same mapping σ can be expressed as a right multiplication by the equation $\sigma(X) = X\rho'$. If $\sigma_i(X) = X\rho_i' \cdot i = 1, 2$, then $\sigma_2(\sigma_1(X)) = (X\rho_1')\rho_2' = X\rho_1'\rho_2'$. From this it follows that the ring of R-homomorphisms of V, that is, the ring of the σ's (with multiplication defined by iteration and addition by linearity), is inverse isomorphic to the ring R' defined above. For this reason we call R' the <u>inverse R-homomorphism ring of</u> V.

Starting with a ring R' and a right R'-space V, we could define the ring of R'-homomorphisms of V as the set of all R'-mappings $X \to X' = (X)\sigma$. But we prefer to use, instead, a left multiplication $\rho''X$ defined by $\rho''X = (X)\sigma$. The set of all mappings ρ'', with addition and multiplication defined in the obvious way, is a ring R'', which we call the <u>inverse</u> R'-<u>homomorphism ring of</u> V.

If, in particular, R' is already the inverse R-homomorphism ring of V, the equation $(\rho X)\rho' = \rho(X\rho')$ states not only that $\rho' \varepsilon R'$, but also that ρ effects an R'-homomorphism ρ'' of V. If R is faithfully represented by its effect on V, we can identify ρ with ρ''. Then we have $R \subseteq R''$.

Starting with R and a "faithful" R-space V, we can set up a sequence of rings R, R', R'', R''', ⋯, each of which is the inverse homomorphism ring of the preceding. R, R'', R''', ⋯ are all left operator domains for V, and R', R''', ⋯ are right domains for V. The relation $R \subseteq R''$ implies in general that $R^{(n)} \subseteq R^{(n+2)}$ We prove that the stronger relations

$$R \subseteq R'' = R^{iv} = \cdots, \quad R' = R''' = \cdots$$

actually hold, by showing that $R' \supseteq R'''$. For let $\rho''' \varepsilon R'''$ be any R''-homomorphism of V. Then, since $R \subseteq R''$, ρ''' is also an R-homomorphism of V, that is, $\rho''' \varepsilon R'$.

We do not yet know any necessary and sufficient conditions on R and V that $R = R''$, but we do have the following theorem.

Main Theorem 5.7A. If R is a semisimple ring faithfully represented by its effect on a space V (= RV which satisfies the minimum condition, then the inverse R-homomorphism ring R' of V is likewise semisimple, and the inverse R'-homomorphism ring R'' of V is equal to R.

In the present section we shall show that the theorem is true for semisimple rings if it is true for simple rings, and then in the following section we shall establish it for simple rings.

The first step in the proof is a theorem on the decomposition of an arbitrary homomorphism into simpler parts, relative to a decomposition of V into a direct sum.

Theorem 5.7B. Let R be any ring, and suppose $V = V_1 + \cdots + V_n$ (direct sum) is an R-space. Any R-homomorphism ρ': $X \rightarrow X' = X\rho'$ of V into itself can be written as a sum $\rho' = \rho_1' + \cdots + \rho_n'$, where ρ_i' is an R-homomorphism of V, which maps V_j into 0 for $j \neq i$. Conversely, any R-homomorphism ρ_i' of V_i into V can be extended into an R-homomorphism of V into V, which maps V_j into 0 for $j \neq i$.

Proof. Suppose $X = X_1 + \cdots + X_n$, $X_i \varepsilon V_i$, is any vector in V. We define ρ_i' by the equation $X\rho_i' = X_i\rho'$. The properties claimed for the ρ_i' are easily checked. As for the converse, we effect the extension by the definition $X\rho_i' = X_i\rho_i'$. This is an R-homomorphism because V_i is an R-space.

Theorem 5.7C. Suppose that $R = \mathfrak{a}_1 + \cdots + \mathfrak{a}_n$ (direct sum) is a semisimple ring with simple ideals $\mathfrak{a}_i = Re_i$. Let V (= RV) be an R-space, and set $V_i = e_iV$. Then $V = V_1 + \cdots + V_n$ (direct sum), and each V_i is mapped into itself by any R-homomorphism ρ' of V into itself; or, in other words, V_i is an R'-space.

Proof. The decomposition of V follows immediately from the orthogonality relations between e's and the equation V = RV, which implies V = 1V (1 is unity element of R). Suppose X_1^0 is a vector in V_1, and let $X_1^0\rho' = X_1 + \cdots + X_n$, where $X_i \in V_i$. Then multiply by e_1, and we get $e_1X_1^0\rho' = X_1^0\rho' = X_1$ (since $e_iV_j = 0$ if $i \neq j$). Hence $X_2 = \cdots = X_n = 0$, or $V_1\rho' \subseteq V_1$. This proves the theorem.

Let $\mathfrak{o}_i'$ denote the inverse R-homomorphism ring of V_i. Since $\mathfrak{o}_jV_i = 0$ if $i \neq j$, $\mathfrak{o}_i'$ is actually the inverse $\mathfrak{o}_i$-homomorphism ring of V_i. It follows from the last two theorems that $R' = \mathfrak{o}_1' + \cdots + \mathfrak{o}_h'$ is a direct decomposition of R' into two-sided ideals. (For we have $\mathfrak{o}_i' = e_i'R' = R'e_i'$, where e_i' is the identity homomorphism on V_i and maps V_j into 0 for $j \neq i$.)

Now, we remark that if V satisfies the minimum condition on subspaces, so does V_i. This reduces the proof of the main theorem to the case where R is simple.

8. THE MAIN RECIPROCITY THEOREM FOR SIMPLE RINGS

Suppose that $V = V_1 + \cdots + V_n$ (direct sum), where each V_i is an irreducible R-space (R any simple ring). By Theorem 5.3E the V_i are all isomorphic as R-spaces. Denote by e_{1i}' an element of R' which effects any (fixed) isomorphism of V_1 onto V_i and which maps $V_2, \cdots, V_n$ into zero. (In particular, let e_{11}' effect the identity mapping of V_1 onto itself.) Let e_{i1}' be an element of R' which maps $V_1, \cdots, V_{i-1}, V_{i+1}, \cdots, V_n$ into zero, and V_i onto V_1 in such a way that $e_{1i}'e_{i1}' = e_{11}'$. Then $e_{ik}' = e_{i1}'e_{1k}'$ effects an isomorphism of V_i onto V_k and maps V_j into zero if $j \neq i$. We verify readily that $e_{ik}'e_{rs}' = \delta_{kr}e_{is}'$ (δ_{kr} (Kronecker's δ-function) is the identity homomorphism if $k = r$ and the zero homomorphism if $k \neq r$).

Since V_1 is irreducible, any nonzero homomorphism of V_1 into itself must just fill V_1 and so has an inverse; in other words, the set of R-homomorphisms of V_1 into itself is a sfield, whose inverse sfield we denote by $\mathfrak{k}'$. Let $\varkappa'$ be an element of $\mathfrak{k}'$, and $X = X_1 + \cdots + X_n$, $X_i \in V_i$, be any element of V. We define $\varkappa'$ as an operator on X_i by the equation $X_i\varkappa' = X_ie_{i1}'\varkappa'e_{1i}'$ and then set $X\varkappa' = X_1\varkappa' + \cdots + X_n\varkappa'$.

<u>Lemma 5.8A</u>. <u>The</u> R-<u>homomorphisms</u> $\varkappa'$ <u>of</u> V <u>commute with the</u> R-<u>homomorphisms</u> e_{ik}' <u>of</u> V.

<u>Proof</u>. We show first that

$$\varkappa' = \Sigma_\nu\, e_{\nu 1}'\varkappa'e_{1\nu}'. \tag{8.1}$$

For $X_i(\Sigma e_{\nu1}'\varkappa'e_{1\nu}') = X_ie_{i1}'\varkappa'e_{1i}' = X_i\varkappa'$, $i = 1, \cdots, n$. Hence the two sides of (8.1) effect the same mapping on V;

therefore they are equal. Now multiplication of (8.1) by e'_{ik} on left and right gives $e'_{ik}\varkappa' = e'_{i1}\varkappa' e'_{1k}$ and $\varkappa' e'_{ik} = e'_{i1}\varkappa' e'_{1k}$, respectively, which establishes the lemma.

It is obvious that $\Sigma e'_{\nu\nu} = 1'$ is the identity R-homomorphism of V into itself. Let $\rho' \varepsilon R'$. Then

$$\rho' = 1'\rho' 1' = \Sigma_{\nu,\mu} e'_{\nu\nu}\rho' e'_{\mu\mu} = \Sigma_{\nu,\mu} \rho'_{\nu\mu},$$

where $\rho'_{ik} = e'_{ii}\rho' e'_{kk}$. Now we wish to show that $\rho'_{ik} = \varkappa' e'_{ik}$ for some $\varkappa' \varepsilon \mathfrak{k}'$. It is clear that ρ'_{ik} maps V_j into zero if $j \neq i$, and that $V_i\rho'_{ik} \subseteq V_k$. The same is true for any $\varkappa' e'_{ik}$. Let $X_i\rho'_{ik} = X_k$ and $X_i e'_{ik} = X'_k$. The mapping $X'_k \to X_k$ is an R-homomorphism of V_k into itself, and so can be effected by (right multiplication by) some element $\varkappa'$ of $\mathfrak{k}'$. Then $\varkappa' e'_{ik} = \rho'_{ik}$, since they effect the same R-homomorphism of V into itself. Thus we have for any $\rho' \varepsilon R'$ a formula

$$(8.2) \qquad \rho' = \Sigma_{\nu,\mu} \varkappa'_{\nu\mu} e'_{\nu\mu}, \text{ where } \varkappa'_{\nu\mu} \varepsilon \mathfrak{k}'.$$

The $\varkappa'_{ik}$ in this formula are unique. For if one of the coefficients, say $\varkappa'_{ik} \neq 0$, then $V_i \rho' e'_{kk} = V_i \Sigma_\nu \varkappa'_{\nu k} e'_{\nu k} = V_i \varkappa'_{ik} e'_{ik} \neq 0$, and so $\rho' \neq 0$. This shows that the expression for 0 is unique; from this it follows that the coefficients in (8.2) are unique for any ρ'. Conversely, it is clear that, for any elements $\varkappa'_{ik}$ of $\mathfrak{k}'$, the formula (8.2) does indeed define an R-homomorphism of V into itself. Thus we have proved that R' _is isomorphic to the ring_ $M_n(\mathfrak{k}')$ _of all n-rowed square_ $\mathfrak{k}'$ _matrices_. Thus R' is a simple ring; this proves the first statement of the main theorem.

Theorem 5.8B. _Let_ R _be a ring with unity element. The_ R-_homomorphisms of_ R _(considered as a left_ R-_space) are just the right multiplications by elements of_ R.

Proof. We have to show that, when V = R, then R' = R also. Let $\xi\rho' = \xi'$ be an R-homomorphism of R into itself. Then $\alpha\xi\rho' = \alpha\xi'$ for any α in R. Take $\xi = 1$ and suppose $1\rho' = 1' = \beta$. Then $\alpha\rho' = \alpha\beta$ for all α in R; that is, ρ' is just right multiplication by β. If we identify ρ' with β, then R' = R. Even without this identification we have that R is isomorphic to R'.

Corollary 5.8C. _If a simple ring_ R _is considered as a left_ R-_space, then its inverse ring_ R' _of_ R-_homomorphisms is again_ R.

This is a new proof of the part of Wedderburn's theorem that states the isomorphism of a simple ring with a total matrix ring. For we have just seen (the statement preceding Theorem 5.8B) that, whenever R is simple, R' is a total matrix ring. Moreover, in the case of the corollary $R = R''$ since they are both equal to R'.

Theorem 5.8D. Let R be a simple ring and V ($\doteq$ RV) an irreducible R-space. Then $R'' = R$.

Proof. We have seen above that the inverse R-homomorphism ring R' of an irreducible R-space V is a sfield $\mathfrak{k}'$. Suppose that V is of $\mathfrak{k}'$-dimension m, that is, $V = X_1 \mathfrak{k}' + \cdots + X_m \mathfrak{k}'$. (Wedderburn's theorem implies that V has finite $\mathfrak{k}'$-dimension.) Then the inverse $\mathfrak{k}'$-homomorphism ring R'' of V is $M_m(\mathfrak{k}')$. But, by Wedderburn's theorem and Theorem 5.6A, $R = M_m(\mathfrak{k}')$. Hence $R = R''$.

We are now ready for the general case again. We suppose R simple and that $V = V_1 + \cdots + V_n$ (direct sum), where each V_i is an irreducible R-space. We proved above that $R' = M_n(\mathfrak{k}')$, where $\mathfrak{k}'$ is the inverse sfield of R-homomorphisms of any V_i. We claim that $\rho'' V_1 \subseteq V_1$. For a vector $X \in V_1$ if and only if $Xe'_{jj} = 0$, $j = 2, \cdots, n$. (For if $X = X_1 + \cdots + X_n$, then $Xe'_{11} = X_1 e'_{11} = X_1$, but if $X \in V_1$, then $Xe'_{11} = X$.) Suppose, now, $X_1 \in V_1$. Then, for $j > 1$, $(\rho'' X_1)e'_{jj} = \rho''(X_1 e'_{jj}) = 0$, since ρ'' is an R'-homomorphism of V. Similarly, $\rho'' V_i \subseteq V_i$. Now $\rho''(X_1 \varkappa') = (\rho'' X_1)\varkappa'$. Hence ρ'' effects a $\mathfrak{k}'$-homomorphism of V_1 into itself. But any $\mathfrak{k}'$-homomorphism of V_1 into itself can be effected by an element of R. Suppose, for instance, that $\rho'' X_1 = \alpha X_1$ for all $X_1 \in V_1$. Now multiplication by e'_{1k} gives $\rho'' X_1 e'_{1k} = \rho'' X_k = \alpha X_k$. Therefore $\rho'' X = \alpha X$ for any $X \in V$; in other words, $\rho'' = \alpha$. (For ρ'' is defined by its effect on V.) This shows that $R'' \subseteq R$, which, together with the universal relation $R \subseteq R''$, gives $R = R''$ and completes the proof of the main theorem (5.7A).

9. A UNIVERSAL MODEL FOR VECTOR SPACES OVER A SIMPLE RING

In the preceding section we had reached the following situation: R is a simple ring, $V = V_1 + \cdots + V_n$ (direct sum), where each V_i is an irreducible (left) R-space. The inverse R-homomorphism ring R' is the total matrix ring $M_n(\mathfrak{k}')$, where $\mathfrak{k}'$ is the inverse R-homomorphism sfield of, say, V_1. Then we saw finally that $R = M_m(\mathfrak{k})$, where $\mathfrak{k}$ is a sfield isomorphic to $\mathfrak{k}'$ and m is the $\mathfrak{k}'$-dimension of V_1. It follows by the main theorem (5.7A) that $V = W_1 + \cdots + W_m$ (direct sum), where each W_i is an irreducible (right) R'-space of $\mathfrak{k}$-dimension n.

Lemma 5.9A. $V_i \cap W_j \neq 0$.

Proof. Suppose $X = X_1 + \cdots + X_n \in W_j$ where, say, $X_k \neq 0$. Then, since W_j is an R'-space, $Xe'_{ki} = X_k e'_{ki} \neq 0$ is in W_j, and also in V_i (since $Ve'_{ki} = V_i$).

Let X_{11} be any nonzero vector in $V_1 \cap W_1$. Then, since the V's and W's are left $\mathfrak{k}$-spaces and right $\mathfrak{k}'$-spaces, $\mathfrak{k}X_{11}\mathfrak{k}' \subseteq V_1 \cap W_1$. Suppose that $\rho X_{11} \in V_1 \cap W_1$. Then $W_1 \supseteq (\rho X_{11})R' = \rho(X_{11}R') = \rho W_1$, that is, ρ transforms W_1 like an element of $\mathfrak{k}$. A similar argument on the right shows that, if $\rho X_{11}\rho' \in V_1 \cap W_1$, then $\rho X_{11}\rho' = \varkappa X_{11}\varkappa'$ for some $\varkappa$ and $\varkappa'$; and therefore shows that $\mathfrak{k}X_{11}\mathfrak{k}' = V_1 \cap W_1$.

Lemma 5.9B. $\mathfrak{k}X_{11} = X_{11}\mathfrak{k}' = V_1 \cap W_1$.

Proof. $RX_{11} = V_1 \supseteq V_1 \cap W_1$. Hence any element of $V_1 \cap W_1$ is of the form ρX_{11}, where $\rho \in R$. But we have just seen that, if $\rho X_{11} \in V_1 \cap W_1$, then there is a $\varkappa$ in $\mathfrak{k}$ for which $\varkappa X_{11} = \rho X_{11}$. (There is also a counting proof which uses the fact that $V_1 = V_1 \cap (W_1 + \cdots + W_m)$ has $\mathfrak{k}$-dimensions m.)

Now we are ready to identify elements in the isomorphic sfields $\mathfrak{k}$ and $\mathfrak{k}'$. We identify $\varkappa$ and $\varkappa'$ if $\varkappa X_{11} = X_{11}\varkappa'$. This identification is justified by showing that the mapping $\varkappa \to \varkappa'$ is an isomorphism. The linearity of the mapping is evident. The 1-1-ness follows by the $\mathfrak{k}$- and $\mathfrak{k}'$-irreducibility of $V_1 \cap W_1$. For multiplication we have $\varkappa_1\varkappa_2 X_{11} = \varkappa_1 X_{11}\varkappa_2' = X_{11}\varkappa_1'\varkappa_2'$, which shows that $(\varkappa_1\varkappa_2)' = \varkappa_1'\varkappa_2'$. (Observe that we are not saying that $\varkappa X = X\varkappa'$ for all X in $V_1 \cap W_1$ if $\varkappa \to \varkappa'$. It is only for X_{11} that we claim this property. Indeed, if instead of X_{11}, we start with cX_{11}, then $\varkappa(cX_{11}) = X_{11}\varkappa'c' = cX_{11}(c^{-1})'\varkappa'c'$ gives a new isomorphism $\varkappa \to (c^{-1})'\varkappa'c'$, which differs from the old by an inner automorphism of $\mathfrak{k}'$. It is only when $\mathfrak{k}$ is commutative that the isomorphism is independent of the choice of X_{11}.)

Now, having identified $\varkappa$ and $\varkappa'$, we have simply $\varkappa X_{11} = X_{11}\varkappa$. Then by the commutativity of $\mathfrak{k}$ and $\mathfrak{k}'$ with the e_{ik} and e'_{ik}, respectively, we have from this that $\varkappa X_{ik} = X_{ik}\varkappa$, where $X_{ik} = e_{k1}X_{11}e'_{1i}$. It is clear that $X_{ik} \in V_i \cap W_k$. We have $V = V \cap V = (V_1 + \cdots + V_n) \cap (W_1 + \cdots + W_m) = \Sigma_{\nu,\mu} V_\nu \cap W_\mu = \Sigma_{\nu,\mu}\mathfrak{k}X_{\nu\mu} = \Sigma_{\nu,\mu} X_{\nu\mu}\mathfrak{k}$. Furthermore, $V_i = \mathfrak{k}X_{i1} + \cdots + \mathfrak{k}X_{im}$; $W_k = \mathfrak{k}X_{1k} + \cdots + \mathfrak{k}X_{nk}$ (both sums direct). Let V_{ik} be the set of all m by n $\mathfrak{k}$-matrices with zeros everywhere except in the intersection of the i-th row and k-th column. Then we have as a universal model for R, V, R' the matrix form:

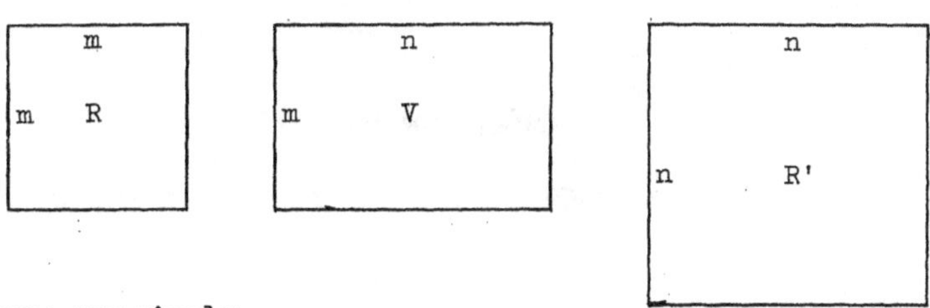

or, more precisely,

$$A_\rho X B_{\rho'},$$

where $A_\rho \varepsilon M_m(\mathfrak{k})$, $B_{\rho'} \varepsilon M_n(\mathfrak{k})$, and X is an m by n $\mathfrak{k}$-matrix.

Another model for R, V and the actual ring R'* of R-homomorphisms expresses the same facts as the one just given and is sometimes more convenient to use, since in it R and R'* are both realized as left operators on V. We use the same vectors X_{ik}, but now arranged in the single column $\begin{Vmatrix} X_{11} \\ \vdots \\ X_{1m} \\ \vdots \\ X_{nm} \end{Vmatrix}$ instead of the m by n array used above.

Then R takes the form

$$\rho \to A'_\rho = \begin{Vmatrix} A_\rho & & & \\ & A_\rho & & 0 \\ & 0 & \ddots & \\ & & & A_\rho \end{Vmatrix} = I_n \times A_\rho \text{ (Kronecker product)},$$

where the m-rowed square matrix A_ρ appears n times, and is the same as the A_ρ given above. Instead of $B_{\rho'}$ we have

$$\rho'^* \to B'_{\rho'^*} = (B_{ik}) = (\varkappa^*_{ik}) \times I_m,$$

where B_{ik} is an m-rowed scalar matrix $\begin{Vmatrix} \varkappa^*_{ik} & & \\ & \ddots & \\ & & \varkappa^*_{ik} \end{Vmatrix}$, $\varkappa^*_{ik}$ being an element of the inverse sfield $\mathfrak{k}^*$ of $\mathfrak{k}$. Observe that A'_ρ and $B'_{\rho'^*}$ are commutative matrices. (For a fuller treatment along these lines see H. Weyl, <u>The Classical Groups</u>, Chapter III.)

CHAPTER VI

KRONECKER PRODUCTS OF SPACES AND RINGS

1. INTRODUCTION

For this chapter we adopt a slightly different notation:

$\mathfrak{k}$, any ring with unit element 1;
α, β, γ, $\varkappa$, elements of $\mathfrak{k}$;
V, a vector space with $\mathfrak{k}$ as right operator domain;
A, A_i, elements of V;
W, a vector space with $\mathfrak{k}$ as left operator domain:
B, B_i, elements of W.

The unit element of $\mathfrak{k}$ is to act as a unit on both V and W. (In later sections of the chapter we shall assume further that V and W are rings with various additional properties.)

In the next section we shall define the Kronecker product space $X = V \times_{\mathfrak{k}} W$. Heuristically speaking, X is the set of all vectors $C = \Sigma A_\nu B_\nu$, where distributivity is assumed with respect to both the A's and the B's, and for many cases this heuristic definition would be satisfactory. In §§2-4 the Kronecker product space is defined with far more generality than is needed for the later applications to the theory of simple rings. In §5 a short cut is indicated for the cases actually used below. (For this short cut see also E. Artin and G. Whaples, "The Theory of Simple Rings," Amer. Journ. Math., 65 (1943), 87-107.)

2. THE KRONECKER PRODUCT OF TWO SPACES

Consider the set Y of all formal (finite) sums $\Sigma A_\nu B_\nu$, with A_i in V and B_i in W. Equality of two such sums is to mean termwise identity (save possibly for order of arrangement); that is, $\Sigma A_\nu B_\nu = \Sigma A'_\mu B'_\mu$ if and only if each term $A_i B_i$ appears as many times on the left side as on the right. The sum of $\Sigma A_\nu B_\nu$ and $\Sigma A'_\mu B'_\mu$ is defined to be $A_1 B_1 + \cdots + A_r B_r + A'_1 B'_1 + \cdots + A'_s B'_s$. (Note that subtraction is not defined, and we do not even assume that $00 = 0$.)

It is our purpose to obtain the Kronecker product $X = V \times_{\mathfrak{k}} W$ as a group whose elements are residue classes in Y. We wish to have the residue class which serves as zero element in X contain all formal sums which would become zero if we assumed distributivity in the A's and B's. As a first step in this direction, we define M as the set of all formal sums which become zero if we assume distributivity in the B's. Stated precisely, M is the set of all formal sums $C = \Sigma A_\nu B_\nu$, for which there is a substitution

$$(2.1)\qquad B_i = \Sigma \beta_{i\nu} B'_\nu, \quad \beta_{ik} \text{ in } \mathfrak{k}$$

such that

$$(2.2)\qquad \Sigma A_\nu \beta_{\nu k} = 0, \quad k = 1, 2, \cdots.$$

(If a B appears more than once, it need not always be expressed in the same way in terms of the B'.)

We establish several properties of the set M. (Denote A(-1) by -A and (-1)B by -B.)

Lemma 6.2A.

I. M is closed under addition.

II. If $\Sigma A_\nu B_\nu$ is in M, then $\Sigma(-A_\nu)B_\nu$ and $\Sigma A_\nu(-B_\nu)$ are also in M.

III. (i). AB + (-A)B is in M;
(ii). 0B is in M;
(iii). A0 is in M.

Proof of I. Let $B_i = \Sigma \beta_{i\nu} B''_\nu$ and $B'_j = \Sigma \beta'_{j\mu} B''_\mu$ be substitutions which show that $C = \Sigma A_\nu B_\nu$ and $C' = \Sigma A'_\mu B'_\mu$ belong to M. Then the same substitutions, taken together, show that C + C' belongs to M. The proof of II is similar. For III (i) we write AB + (-A)B in the form $AB_1 + (-A)B_2$, and then the substitution $B_1 = 1B$, $B_2 = 1B$ gives A + (-A) = 0. The substitutions B = 1B and 0 = 0B establish III (ii) and III (iii), respectively.

The set M is not quite large enough to serve as the zero residue class. For M might contain C' and C + C' without containing C, and we shall see below that this would interfere with setting up an effective residue-class calculus. To get around this difficulty we define a second subset N of Y by the statement

(2.3) $C \in N$ if and only if there is an element C' of Y such that $C + C' + C'_- \in M$.

(If $C = \Sigma A_\nu B_\nu$, we denote $\Sigma(-A_\nu)B_\nu$ by C_-.) It is obvious that the preceding lemma holds with N in place of M.

We have, in addition:

Lemma 6.2A'.

IV. $M \subseteq N$.

V. *If* C' *and* $C + C'$ *belong to* N, *then* C *belongs to* N.

To establish IV note that $C \in M$ implies $C + AB + (-A)B \in M$ by III and I of Lemma 6.2A.

Next suppose C' and $C + C'$ in N, or, from (2.3), that

$$C' + C'' + C''_- \in M, \tag{2.4}$$

and

$$C + C' + C''' + C'''_- \in M. \tag{2.5}$$

Now, applying Lemma 6.2A II to the sum in (2.4), we see that $C'_- + C''_- + C'' \in M$. Add this to (2.5) (remembering that M is closed under addition) and we get, after rearranging terms:

$$C + (C' + C'' + C''') + (C'_- + C''_- + C'''_-) \in M$$

or $C \in N$, which is just the contention in V.

We say that $C \equiv C'$ (mod N) (read "C is congruent to C' mod N") if $C + C'_- \in N$.

Lemma 6.2B. Congruence mod N *is* (i) *reflexive,* (ii) *symmetric,* (iii) *transitive,* and (iv) *additive.*

Proof. Parts (i) and (ii) are obvious and are true even in the smaller set M. For (iii) we suppose $C \equiv C'$ (mod N) and $C' \equiv C''$ (mod N). Then $(C + C'_-) \in N$ and $(C' + C''_-) \in N$ by the definition of congruence. Add these two elements, and we get $C + C''_- + (C' + C'_-) \in N$. By Lemma 6.2A III (i), $C' + C'_- \in N$, and so $C + C''_- \in N$ by part V of the preceding lemma. Hence $C \equiv C''$, that is, *the congruence relation is transitive.* Next, if $C \equiv C'$ and $C'' \equiv C'''$, we have $C + C'_-$ and $C'' + C'''_-$ in N and, therefore, also $(C + C'') + (C'_- + C'''_-)$ in N; that is, $C + C' \equiv C'' + C'''$, which proves that *the congruence relation is additive.*

We now define the sum of two residue classes to be the residue class containing an element obtained by adding any element of the first residue class to any element of the second residue class. The following theorem is an immediate corollary to the lemma just proved.

Theorem 6.2C. The set of residue classes mod N *in the set* Y *of all formal sums* $\Sigma A_\nu B_\nu$ *is a group* X *under* addition, the identity class being N itself, the inverse of the class containing $C = \Sigma A_\nu B_\nu$ being the class containing $C_- = \Sigma(-A_\nu)B_\nu$.

This group $X = V \times_{\mathfrak{t}} W$ is called the Kronecker product of V and W relative to $\mathfrak{t}$.

This theorem justifies us in considering X as the set of all formal (finite) sums $C = \Sigma A_\nu B_\nu$, with equality now defined by

$$C = C' \text{ if } C + C'_- \varepsilon N. \tag{2.6}$$

From now on we shall use the equality sign in this new sense only. Whenever we write $C = C'$ we are thinking of C, C' as elements of X, but when we say $C \varepsilon N$ we are thinking of C as an element of Y.

3. THE SYMMETRY OF A KRONECKER PRODUCT IN ITS FACTORS

Although our definitions of M and N and, consequently, of $X = V \times_{\mathfrak{t}} W$ were not symmetric in V and W, it is nevertheless true that V and W play completely analogous roles in X. To indicate that our original M, as defined by (2.1) and (2.2), involves only substitutions on the vectors of W we shall henceforth designate it by M_W; and, likewise, the original N of (2.3) we now designate by N_W. Next, we introduce M_V as the set of elements $\Sigma A_\nu B_\nu$ in Y, for which there exists a substitution

$$A_k = \Sigma A'_\nu \alpha_{\nu k}, \quad \alpha_{ik} \text{ in } \mathfrak{t} \tag{3.1}$$

such that

$$\Sigma \alpha_{i\nu} B_\nu = 0 \quad i = 1, 2, \cdots \tag{3.2}$$

Then we define N_V by the statement that

(3.3) $\Sigma A_\nu B_\nu \varepsilon N_V$ if and only if there is an element $\Sigma A'_\mu B'_\mu$ of Y such that $\Sigma A_\nu B_\nu + \Sigma A'_\mu B'_\mu + \Sigma A'_\mu(-B'_\mu) \varepsilon M_V$.

It is clear that M_V, N_V have all the essential properties of M_W, N_W, so that we could introduce a group of residue classes mod N_V in Y. However, the following theorem shows that this group would just be our original X again, and so it establishes the symmetry of the Kronecker product in its factors.

Theorem 6.3A. $N_V = N_W$.

Before proving this theorem we obtain several lemmas, which, in addition to aiding in the proof of the theorem, embody important properties of the Kronecker product group.

Lemma 6.3B.

(i). $A(B_1 + B_2) = AB_1 + AB_2$;

(ii). $(A_1 + A_2)B = A_1B + A_2B$;

(iii). $A(\beta B) = (A\beta)B$.

(Note that the equality here is the new equality defined in the last paragraph of §2.)

Lemma 6.3B'.

(i). $A(B_1 + B_2) + (-A)B_1 + (-A)B_2 \in M_W$;

(ii). $(A_1 + A_2)B + (-A_1)B + (-A_2)B \in M_W$;

(iii). $A(\beta B) + (-A\beta)B \in M_W$.

Since $M_W \subseteq N_W$, the first of these lemmas will follow from the second. To establish the second lemma we list suitable substitutions of the type (2.1) for each part:

(i) $(B_1 + B_2) = 1B_1 + 1B_2$, $B_1 = 1B_1 + 0B_2$, $B_2 = 0B_1 + 1B_2$;

(ii) $B = 1B$, $B = 1B$, $B = 1B$; (iii) $(\beta B) = \beta B$ and $B = 1B$.

Lemma 6.3C. $M_V \subseteq N_W$.

Proof. Suppose $A_k = \Sigma A'_\nu \alpha_{\nu k}$ and $\Sigma\alpha_{i\nu}B_\nu = 0$, $i = 1, 2, \cdots$; or, in other words, that $\Sigma A_\nu B_\nu \in M_V$. Then we have

$$\begin{aligned}\Sigma A_\nu B_\nu &= \sum_\nu(\sum_\mu A'_\mu\alpha_{\mu\nu})B_\nu \\ &= \sum_{\nu,\mu}(A'_\mu\alpha_{\mu\nu})B_\nu \text{ (by Lemma 6.3B (ii))} \\ &= \sum_{\nu,\mu}A'_\mu(\alpha_{\mu\nu}B_\nu) \text{ (by Lemma 6.3B (iii))} \\ &= \sum_\mu A'_\mu(\sum_\nu\alpha_{\mu\nu}B_\nu) \text{ (by Lemma 6.3B (i))} \\ &= \sum_\mu A'_\mu 0 = 0;\end{aligned}$$

that is, $\Sigma A_\nu B_\nu \in N_W$, which is what the lemma claims.

Now we are ready to prove the theorem. Let $\Sigma A_\nu B_\nu \in N_V$. Then there is an element $\Sigma A'_\mu B'_\mu$ of Y such that $\Sigma A_\nu B_\nu + [\Sigma A'_\mu B'_\mu + \Sigma A'_\mu(-B'_\mu)] \in M_V$. Clearly the bracketed part belongs to M_W, and therefore to N_W. Now since $M_V \subseteq N_W$ we have $\Sigma A_\nu B_\nu = 0$ by Lemma 6.2A'; hence $N_V \subseteq N_W$. Now repeat the same arguments, with the roles of V and W interchanged, to get $N_W \subseteq N_V$ and hence $N_W = N_V$, as was to be proved.

4. FURTHER PROPERTIES OF M_W

Suppose that the substitution $B_i = \Sigma\beta_{i\nu}B'_\nu$ yields

(4.1) $$\Sigma A_\nu\beta_{\nu k} = 0, \quad k = 1, 2, \cdots.$$

Suppose, further, that $B'_i = \Sigma\gamma_{i\nu}B''_\nu$. Then, combining the two substitutions, we get $B_i = \Sigma\alpha_{i\nu}B''_\nu$, where $\alpha_{ik} = \Sigma\beta_{i\nu}\gamma_{\nu k}$. Now

(4.2) $$\Sigma A_\nu\alpha_{\nu k} = \sum_\nu A_\nu\sum_\mu\beta_{\nu\mu}\gamma_{\mu k} = \sum_\mu\sum_\nu(A_\nu\beta_{\nu\mu})\gamma_{\mu k} = \sum_\mu 0\gamma_{\mu k} = 0.$$

Both (4.1) and (4.2) show that $\Sigma A_\nu B_\nu \in M_W$. This proves

Lemma 6.4A. If a substitution shows that an element C of Y is in M_W, then any product of substitutions of which the given substitution is the first factor will also show that C is in M_W.

For the next several sections we shall need a stronger kind of independence than that defined in Chapter I, §4. We shall now say that the vectors $B_1, \cdots, B_r$ are independent if $\beta_1B_1 + \cdots + \beta_rB_r = 0$ requires $\beta_1 = \cdots = \beta_r = 0$. Note that, if $\mathfrak{k}$ is a sfield, this definition is equivalent to the old one.

Lemma 6.4B. Let $W = \mathfrak{k}B''_1 + \cdots + \mathfrak{k}B''_r$ where $B''_1, \cdots B''_r$ are independent, that is, the sum is direct). Suppose $B_i = \Sigma\alpha_{i\nu}B''_\nu$, $i = 1, 2, \cdots$. Then $\Sigma A_\nu B_\nu \in M_W$ if and only if $\Sigma A_\nu\alpha_{\nu k} = 0$, $k = 1, \cdots, r$. Furthermore, $N_W = M_W$.

Proof. The "if" is just the definition of M_W. As to the "only if," suppose $B_i = \Sigma\beta_{i\nu}B'_\nu$ and $\Sigma A_\nu\beta_{\nu k} = 0$. Since $B''_1, \cdots, B''_r$ are a $\mathfrak{k}$-basis for W, there exists a substitution of the form $B'_i = \Sigma\gamma_{i\nu}B''_\nu$. By the preceding lemma we have, then, $\Sigma A_\nu\alpha'_{\nu k} = 0$, where $B_i = \Sigma\alpha'_{i\nu}B''_\nu$ is the product substitution. But since the B''_i are linearly independent, $\alpha'_{ik} = \alpha_{ik}$, and so the first part of the lemma is proved.

Next, let $\Sigma A_\nu B_\nu$ be any element of N_W. Then by the definition (2.3) of N_W we have an element $\Sigma A'_\mu B'_\mu$ of Y such that $\Sigma A_\nu B_\nu + \Sigma A'_\mu B'_\mu + \Sigma(-A'_\mu)B'_\mu \in M_W$. If we express all the B_i and B'_i in terms of the B''_i we shall get $\Sigma A_\nu\alpha_{\nu k} + \Sigma A'_\mu\beta_{\mu k} + \Sigma(-A'_\mu)\beta'_{\mu k} = 0$, $k = 1, 2, \cdots$. But since the B''_i are independent we must have $\beta_{ik} = \beta'_{ik}$, and hence $\Sigma A_\nu\alpha_{\nu k} = 0$. This shows that $\Sigma A_\nu B_\nu \in M_W$, that is, $N_W \subseteq M_W$. We know already (Lemma 6.2A') that $M_W \subseteq N_W$, and so we conclude that $N_W = M_W$.

We say that the space W is locally finite if every finite set of vectors in W is in a subspace generated by a finite number of independent vectors. (If $\mathfrak{k}$ is a sfield V and W are always locally finite.)

Lemma 6.4C. If W is locally finite, then $N_W = M_W$. Furthermore, if $B_1, B_2, \cdots$ lie in the subspace W'' of W spanned by the independent vectors $B_1'', \cdots, B_r''$, say $B_i = \Sigma\alpha_{i\nu}B_\nu''$, then $\Sigma A_\nu B_\nu \in M_W$ if and only if $\Sigma A_\nu \alpha_{\nu k} = 0$, $k = 1, \cdots, r$.

This lemma follows from the formulas used in proving the preceding lemma if we operate in suitable finite subspaces of W. In any given part of the argument only a finite set S of vectors of W is involved. To carry out this given part of the argument, we choose any subspace $W' = \mathfrak{k}B_1'' + \cdots + \mathfrak{k}B_r'' + \cdots + \mathfrak{k}B_s''$ which contains all the vectors of S and then apply the corresponding formulas of the proof of Lemma 6.4B, with W' in place of the "W" used there.

5. KRONECKER PRODUCTS OF LOCALLY FINITE SPACES

In this section we indicate a short cut to the theory of the Kronecker product of two locally finite spaces. The general plan of this short cut would be

(1). Lemma 6.2A;
(2). Lemma 6.4A;
(3). Part V of Lemma 6.2A' for M_W;
(4). Lemma 6.2B for congruences mod M_W;
(5). Theorem 6.2C, with "mod M_W" instead of mod N_W";
(6). Lemma 6.3B;
(7). Lemma 6.4C, second part.

Then to establish the symmetry of the Kronecker product in its factors one would show that $M_V = M_W$. Note that the sets N_V, N_W are not used at all in this development. The only steps of this short cut which require new proofs (that is, other than those already given above) are (3) and the statement $M_V = M_W$. These gaps in the proof are supplied by the following lemma. (This short cut is carried out completely in the paper previously cited, "The Theory of Simple Rings," by E. Artin and G. Whaples.)

Lemma 6.5A. If W is locally finite, $M_V \subseteq M_W$; furthermore, if C' and C + C' are in M_W, then C is in M_W.

Proof. Suppose that the substitution $A_k = \Sigma A'_\nu \alpha_{\nu k}$ gives

$$(5.1) \qquad \Sigma \alpha_{i\nu} B_\nu = 0, \quad i = 1, 2, \cdots;$$

that is, $\Sigma A_\nu B_\nu \in M_V$. Let $B'_1, \cdots, B'_r$ be independent vectors in terms of which $B_1, B_2, \cdots$ can be expressed, for example,

$$(5.2) \qquad B_i = \Sigma \beta_{i\nu} B'_\nu .$$

Substituting this in (5.1), we get $\Sigma \alpha_{i\nu} \beta_{\nu\mu} B'_\mu = 0$, $i = 1, 2, \cdots$; or, since the B'_i are independent,

$$(5.3) \qquad \Sigma \alpha_{i\nu} \beta_{\nu k} = 0, \quad i = 1, 2, \cdots, \quad k = 1, \cdots, r.$$

Now substitute for $B_1, B_2, \cdots$ the expressions (5.2), and then from $\Sigma A_\nu \beta_{\nu k} = \Sigma A'_\mu \alpha_{\mu\nu} \beta_{\nu k} = 0$ (by (5.3)) we conclude that $\Sigma A_\nu B_\nu \in M_W$. But $\Sigma A_\nu B_\nu$ was any element of M_V, and so this proves that $M_V \subseteq M_W$.

Next suppose that C' and $C + C' \in M_W$, where $C = \Sigma A_\nu B_\nu$ and $C' = \Sigma A'_\mu B'_\mu$. Let $B''_1, \cdots, B''_r$ be independent vectors such that

$$(5.4) \qquad B_i = \Sigma \beta_{i\nu} B''_\nu, \quad B'_i = \Sigma \beta_{i\mu} B''_\mu .$$

Since C' and $C + C' \in M_W$, we have by Lemma 6.4C that

$$\Sigma A'_\mu \beta'_{\mu k} = \Sigma A_\nu \beta_{\nu k} + \Sigma A'_\mu \beta'_{\mu k} = 0, \quad k = 1, \cdots, r.$$

Now by subtraction we get $\Sigma A_\nu \beta_{\nu k} = 0$, $k = 1, \cdots, r$, that is, $C \in M_W$. This completes the proof of the lemma.

If, now, we suppose further that V is locally finite, the same arguments with the roles of W and V reversed give $M_W \subseteq M_V$, and therefore $M_V = M_W$.

6. EXAMPLES OF KRONECKER PRODUCTS

Let $\mathfrak{k}$ be the ring of integers, V the field of rational numbers, and W the field of residue classes (of the integers) modulo any prime number p. Then $V \times_{\mathfrak{k}} W = 0$. For $AB = (A/p \cdot p)B = (A/p)(pB) = (A/p)0 = 0$.

If we take $\mathfrak{k}$ and W as above and let V be the integers, then $V \times_{\mathfrak{k}} W = W$. For in this case $V = 1\mathfrak{k}$ is finite. Any element A in V is also $1A$ (where A is now thought of as in $\mathfrak{k}$). Hence $\Sigma A_\nu B_\nu = \Sigma (1_\nu A_\nu) B_\nu = \Sigma 1_\nu (A_\nu B_\nu) = 1(\Sigma A_\nu B_\nu)$, that is, $V \times_{\mathfrak{k}} W = 1W$, and if we identify $1B$ with B, we have $V \times_{\mathfrak{k}} W = W$ as claimed.

7. THE KRONECKER PRODUCT AS A SPACE

If, in addition to our previous hypotheses on V, $\mathfrak{k}$, and W, we suppose that V is a left R-space (R any ring) such that $a(A\varkappa) = (aA)\varkappa$, then we can make $X = V \times_{\mathfrak{k}} W$ into a left R-space by the definition

$$(7.1) \qquad a\Sigma A_\nu B_\nu = \Sigma(aA_\nu)B_\nu.$$

To justify this definition we must show that the subset N of Y is closed under left multiplication by elements of R. (Having proved that $N_V = N_W$, we henceforth denote both by N.) Let $\Sigma A_\nu B_\nu \in M_W$. Then we have a substitution $B_i = \Sigma\beta_{i\nu}B_\nu'$ such that $\Sigma A_\nu\beta_{\nu k} = 0$, $k = 1, 2, \cdots$. Then, from $\Sigma aA_\nu\beta_{\nu k} = 0$, we conclude that $\Sigma(aA_\nu)B_\nu \in M_W$, that is, $RM_W \subseteq M_W$. Next, suppose that $\Sigma A_\nu B_\nu$ is any element of N and that $\Sigma A_\nu B_\nu + \Sigma A_\mu' B_\mu' + \Sigma(-A_\mu')B_\mu' \in M_W$. Then, from $\Sigma(aA_\nu)B_\nu + \Sigma(aA_\mu')B_\mu' + \Sigma(-aA_\mu')B_\mu' \in M_W$, we conclude that $\Sigma(aA_\nu)B_\nu \in N$, that is, $RN \subseteq N$. This completes the justification of definition (7.1).

In the same way we can show, if W is a right R'-space, that $V \times_{\mathfrak{k}} W$ can be made into a right R'-space.

If W is a left R-space, we might attempt to make $V \times_{\mathfrak{k}} W$ into a left R-space by the definition

$$(7.2) \qquad a\Sigma A_\nu B_\nu = \Sigma A_\nu(aB_\nu).$$

As above, we would have to justify this definition by showing that

$$(7.3) \qquad \Sigma A_\nu B_\nu \in M_W \text{ implies } \Sigma A_\nu(aB_\nu) \in M_W.$$

It is easy to see that (7.3) does not hold in general. A sufficient condition for the validity of (7.3) is

$$(7.4) \qquad a(\varkappa B) = \varkappa(aB) \text{ for all } a \in R,\ \varkappa \in \mathfrak{k},\ B \in W.$$

For let $B_i = \Sigma\beta_{i\nu}B_\nu'$ give $\Sigma A_\nu\beta_{\nu k} = 0$, $k = 1, 2, \cdots$. Then the substitution $aB_i = \Sigma a\beta_{i\nu}B_\nu' = \Sigma\beta_{i\nu}(aB_\nu')$ shows that $\Sigma A_\nu(aB_\nu) \in M_W$.

If W is a left R-space, let R' be inverse isomorphic to R. We make W into a right R'-space by the equation $aB = Ba'$, where a and a' are corresponding elements of R and R'. Observe $(ab)B = B(ab)' = Bb'a'$ checks with the inverse isomorphism of R and R'; and is, of course, the reason for taking R' inverse isomorphic to R. The above condition $a(\varkappa B) = \varkappa(aB)$ connecting elements of $\mathfrak{k}$ and R is equivalent to the condition $\varkappa(Ba') = (\varkappa B)a'$ between $\mathfrak{k}$ and R'. Thus the associative law for left and right operator domains is strictly analogous to the commutative law for two left (or right) operator domains.

Let $\mathfrak{k}$ and $\mathfrak{K}$ be two rings, and suppose that V is a right $\mathfrak{k}$ space, W is a left $\mathfrak{k}$ space and a right $\mathfrak{K}$ space (with $(\mathfrak{k}W)\mathfrak{K} = \mathfrak{k}(W\mathfrak{K})$), and that Z is a left $\mathfrak{K}$ space. Then, according to the preceding discussion, $V \times_{\mathfrak{k}} W$ is a right $\mathfrak{R}$ space, and so $X = (V \times_{\mathfrak{k}} W) \times_{\mathfrak{R}} Z$ is defined. Similarly, $Y = V \times_{\mathfrak{k}} (W \times_{\mathfrak{R}} Z)$ is defined. Abstractly speaking, X and Y are, of course, distinct, but it may happen that they are isomorphic. If, in particular, the mapping $\Sigma(A_\nu B_\nu)C_\nu \longleftrightarrow \Sigma A_\nu (B_\nu C_\nu)$, is an isomorphism, we identify X and Y, writing merely $V \times_{\mathfrak{k}} W \times_{\mathfrak{R}} Z$, and say that for these factors the Kronecker product is associative. We state, without proof, that a sufficient condition for this associativity is that V and Z be locally finite. (A proof can be based on Theorem 6.4C.)

8. THE KRONECKER PRODUCT OF TWO RINGS

If V and W are rings, we wish to make their Kronecker product into a ring by a suitable definition of multiplication. We must have a product

$$(8.1) \qquad (\Sigma A_\nu B_\nu)(\Sigma A'_\mu B'_\mu)$$

which is linear in both factors and such that 0 has the multiplicative properties of the zero of a ring.

The following theorem gives sufficient conditions for $X = V \times_{\mathfrak{k}} W$ to be a ring. These conditions are restricted enough for our needs, so that we do not enter into any discussion of the possibility of weakening them.

Theorem 6.8A. Let V and W be rings with $\mathfrak{k}$ as both left and right operator domain for each such that $A\kappa = \kappa A$, $\kappa B = B\kappa$ for all $A \varepsilon V$, $B \varepsilon W$, $\kappa \varepsilon \mathfrak{k}$. If we set

$$(8.2) \qquad (\Sigma A_\nu B_\nu)(\Sigma A'_\mu B'_\mu) = \sum_{\nu,\mu} (A_\nu A'_\mu)(B_\nu B'_\mu),$$

then $X = V \times_{\mathfrak{k}} W$ is again a ring, called the Kronecker direct product ring of V and W relative to $\mathfrak{k}$.

From $A(\kappa_1\kappa_2) = (\kappa_1\kappa_2)A = (A\kappa_1)\kappa_2 = (\kappa_1 A)\kappa_2 = (\kappa_2\kappa_1)A$ we get $(\kappa_1\kappa_2 - \kappa_2\kappa_1)A = 0$ for all $A \varepsilon V$, $\kappa_1, \kappa_2 \varepsilon \mathfrak{k}$, and the same for elements B of W. Hence the set of elements $\kappa_1\kappa_2 - \kappa_2\kappa_1$ (sometimes called commutators) generate a two-sided ideal $\mathfrak{a}$ of $\mathfrak{k}$, which annihilates both V and W. The residue class ring $\mathfrak{k}^* = \mathfrak{k} - \mathfrak{a}$ is commutative and effects the same mappings on V and W as $\mathfrak{k}$. It is clearly sufficient to prove the theorem for $\mathfrak{k}^*$ instead of $\mathfrak{k}$; or, in other words, there is no real loss of generality in assuming $\mathfrak{k}$ commutative in the first place, and we shall do this.

It is clear that the definition of multiplication given in (8.2) satisfies the associative law and is distributive on both sides. It remains to show that, if $C = \Sigma A_\nu B_\nu \in N$, then CC' and $C'C$ both belong to N, for any $C' = \Sigma A'_\mu B'_\mu$. In view of the distributive law it is sufficient to consider only $C' = AB$.

We first check products of an element $C \in M_W$. Suppose, say, that $B_i = \Sigma \beta_{i\nu} B'_\nu$ and $\Sigma A_\nu \beta_{\nu k} = 0$, $k = 1, 2, \cdots$. Note that $(BB_i) = \Sigma B(\beta_{i\nu} B'_\nu) = \Sigma B(B'_\nu \beta_{i\nu})$, since β_{ik} is commutative with B'_k, $= \Sigma (BB'_\nu)\beta_{i\nu}$, since $\mathfrak{k}$ is a right operator domain for the ring W, $= \Sigma \beta_{i\nu}(BB'_\nu)$, since β_{ik} is commutative with BB'_k. Now $\Sigma(AA_\nu)\beta_{\nu k} = \Sigma A(A_\nu \beta_{\nu k}) = 0$, $k = 1, 2, \cdots$ shows that $ABC \in M_W$. Similar steps establish the right closure of M_W. (Observe that left closure of M_W depends only on the relations between $\mathfrak{k}$ and W and the fact that $\mathfrak{k}$ is a right operator domain for the ring V, and that right closure of M_W depends only on the relations between $\mathfrak{k}$ and V and the fact that $\mathfrak{k}$ is a left operator domain for the ring W.)

To complete the proof of the theorem we have to check closure of N under multiplication. This follows easily from the closure of M_W. For let $C + C' + C'_- \in M_W$, that is, $C \in N$. Then $ABC + ABC' + ABC'_- \in M_W$. But, since $(ABC')_- = ABC'_-$, this shows that $ABC \in N$. Closure under right multiplication follows similarly.

9. KRONECKER PRODUCTS OF DIRECT SUMS

We suppose that $\mathfrak{k}$, V, W, satisfy the hypotheses of Theorem 6.8A, so that their Kronecker product $X = V \times_{\mathfrak{k}} W$ is again a ring.

Theorem 6.9A. Suppose $V = V_1 + V_2$ (direct sum). Then $X = X_1 + X_2$ (direct sum) where $X_i = V_i \times_{\mathfrak{k}} W$, $i = 1, 2$.

Proof. Let the relations $B_i = \Sigma \beta_{i\nu} B'_\nu$, $\Sigma A_\nu \beta_{\nu k} = 0$ show that $\Sigma A_\nu B_\nu \in M_W$. Since $V = V_1 + V_2$ is a direct sum we have for each A_i a unique expression of the form $A_i = A_i^1 + A_i^2$. Then, writing $0 = \Sigma A_\nu \beta_{\nu k} = \Sigma A_\nu^1 \beta_{\nu k} + \Sigma A_\nu^2 \beta_{\nu k}$, we conclude from $V_1 \cap V_2 = 0$ that $\Sigma A_\nu^1 \beta_{\nu k} = \Sigma A_\nu^2 \beta_{\nu k} = 0$, that is, $\Sigma A_\nu^i B_\nu \in M_W$, $i = 1, 2$. Using the relationship between M_W and N and this result, we show easily that, if $\Sigma A_\nu B_\nu \in N$, then $\Sigma A_\nu^i B_\nu \in N$ $i = 1, 2$. This proves that $X \supseteq X_i$, $i = 1, 2$ (strictly speaking this proves that X contains a subring isomorphic to X_i, but we can and do identify this subring with X_i) and that $X = X_1 + X_2$. To see that this sum is direct we observe that (again using arguments like those just above) $X_1 \cap X_2 = 0$.

Corollary 6.9B. Suppose $V = A_1\mathfrak{k} + \cdots + A_n\mathfrak{k}$ and $W = \mathfrak{k}B_1 + \cdots + \mathfrak{k}B_m$ (both sums direct), where $\mathfrak{k}$ is a field. Then $V \times_{\mathfrak{k}} W = \sum_{\nu,\mu} A_\nu \mathfrak{k} B_\mu$; and if $\mathfrak{k}$ is in the center of V, the $\mathfrak{k}$-dimension of $V \times_{\mathfrak{k}} W$ is mn.

10. REMARKS ON THE MINIMUM CONDITION IN KRONECKER PRODUCTS

We open this section with an example which shows that the Kronecker product ring $X = V \times_{\mathfrak{k}} W$ does not necessarily have the minimum condition on left ideals, even though both the factors V and W do. We need the following theorem for this example.

Theorem 6.10A. Let R be a ring without null divisors and with minimum condition on left ideals. Then R is a sfield.

This theorem follows at once from the structure theorems for semisimple rings. We give here a direct proof, based on showing that the nonzero elements of R form a group under the ring multiplication. Suppose that a is any nonzero element of R. Then, because of the minimum condition, the descending chain of left ideals R, Ra, $Ra^2, \cdots$ must have all but a finite number of its terms equal. Suppose $Ra^\nu = Ra^{\nu+1}$; this means that the equation $ba^\nu = ca^{\nu+1}$ has a solution c for every b in R. Since there are no zero divisors, this solution is unique. Denote by e the solution when b = a and by a' the solution when b = e. Since R has no zero divisors, cancellation is permitted. Hence a = ea and e = a'a. Multiplication of the first of these equations on the left by a and then cancellation of an a from the right gives a = ae. Now multiplication of this by any element b on the right and cancellation of a from the left gives b = eb. Thus e is a left unit for R, and therefore a' is a left inverse for a. Since a was any nonzero element of R, this shows that the nonzero elements of R form a group and proves that R is a sfield.

Now let $\mathfrak{k}$ be the field of complex numbers, let $V = \mathfrak{k}(x)$ and $W = \mathfrak{k}(y)$, x and y being indeterminants over $\mathfrak{k}$ ($\mathfrak{k}(x)$ denotes the set of all rational functions in x with coefficients in $\mathfrak{k}$). The Kronecker product $X = V \times_{\mathfrak{k}} W$ is contained in the field $\mathfrak{k}(x, y)$. For if $\Sigma f_\nu(x) g_\nu(y) \in M_W$, then $\Sigma f_\nu(x) g_\nu(y) = 0$ in $\mathfrak{k}(x, y)$. Hence X has no zero divisors. The elements of X all have the form $P(x, y)/P_1(x)P_2(y)$, where the P's are $\mathfrak{k}$-polynomials, that is, X is the set of rational functions with "fixed poles." (Such a rational function as $1/(xy + 1)$ is not in X.) Hence X _is_ a proper subring of $\mathfrak{k}(x, y)$, clearly _not a sfield_. The preceding theorem now precludes the possibility of the minimum condition holding in X, even though it is obviously true in both factors V and W.

Although we can say nothing, in general, about the validity of the minimum condition in a Kronecker product, the following theorem gives a sufficient condition in case one of the factors is finite.

Theorem 6.10B. Let V, W, _and_ $\mathfrak{k}$ _satisfy the hypotheses of Theorem 6.8A. Suppose, further, that_ $W = \mathfrak{k}B_1 + \cdots + \mathfrak{k}B_r$ _and that the minimum condition on left ideals holds in both_ V _and_ W. _Then the minimum condition holds in the Kronecker product ring_ $X = V \times_{\mathfrak{k}} W$.

We have $X = VB_1 + \cdots + VB_r$, and so by Theorem 2.2D (with $R = X$ and $R' = V$) the minimum condition on left ideals holds in X.

CHAPTER VII

SIMPLE RINGS CONTINUED*

1. ANALYTIC LINEAR TRANSFORMATION

Let R be a simple ring and $\mathfrak{k}$ its center. $\mathfrak{k}$ is a field; this follows easily from Wedderburn's theorem, but we give here a direct proof. The unit element 1 of R is contained in $\mathfrak{k}$ and, if κ is any nonzero element of $\mathfrak{k}$, then $R\kappa$, being a nonzero two-sided ideal of R, is equal to R. But this means that R contains an element κ^{-1} such that $\kappa^{-1}\kappa = 1$. κ^{-1} is in $\mathfrak{k}$, for, if r is any element of R, then $\kappa^{-1}r = \kappa^{-1}r1 = \kappa^{-1}r\kappa\kappa^{-1} = \kappa^{-1}\kappa r\kappa^{-1} = 1r\kappa^{-1} = r\kappa^{-1}$. But this is sufficient to show that $\mathfrak{k}$ is indeed a field.

A function $\sigma(x)$ of R into itself will be called a $\mathfrak{k}$-linear function if the mapping $x \to \sigma(x)$ is a homomorphism of R, considered as a $\mathfrak{k}$-space. A function $\chi(x)$ which can be written in the form $\chi(x) = \Sigma a_\nu x a'_\nu$, where x runs through R, and the a_ν, a'_ν are elements of R, is called an <u>analytic linear function</u>. Clearly all such functions are included among the $\mathfrak{k}$-linear functions. If χ and χ' are two analytic functions, their product $\chi'' = \chi(\chi'(x))$ is again analytic. The main theorem about analytic functions is the following.

<u>Theorem 7.1A.</u> <u>If</u> R <u>is a simple ring with center</u> $\mathfrak{k}$ <u>and if</u> $a_1, a_2, \cdots, a_r$ <u>are independent with respect to</u> $\mathfrak{k}$, <u>we can find an analytic linear function mapping these elements on arbitrarily chosen elements</u> $b_1, b_2, \cdots, b_r$.

<u>Proof.</u> It obviously suffices to prove the theorem in case one of the b_ν is 1 and the others are 0. If $r = 1$, the possible values $\chi(a_1)$ form the two-sided ideal Ra_1R, which is R because R is simple: that proves our theorem for $r = 1$. Let it be true for $r - 1$, and assume that we have found, for each $i = 1, \cdots, r - 1$, an analytic linear function $\chi_i(x)$ that maps a_i into 1 and all the others (except possibly a_r) into 0. We try to find an $\chi(x)$ which maps a_r into 1 and all the other a_i into 0. This is easily

---

*Arranged by Dr. George Whaples from his paper with Professor Artin, "The theory of simple rings," <u>Amer. Journ. Math.</u>, 65 (1943), 87-107.

done as soon as any $\chi_i(a_r)$ is an element outside the center $\mathfrak{t}$; indeed, if c is not commutative with $\chi_i(a_r)$, the function $p(x) = \chi_i(x)c - c\chi_i(x)$ will map $a_1, a_2, \cdots, a_{r-1}$ into 0, and a_r into a nonzero element b. If $q(x)$ is an analytic linear function mapping b into 1, we can take $q(p(x))$ as our desired function $\chi(x)$.

We therefore assume that $\chi_i(a_r) = \kappa_i \in \mathfrak{t}$ for $i = 1, \cdots, r - 1$. We now put $p(x) = a_1\chi_1(x) + a_2\chi_2(x) + \cdots + a_{r-1}\chi_{r-1}(x) - x$. Obviously, $p(a_i) = 0$ for $i = 1, \cdots, r - 1$. $p(a_r)$ cannot be 0, or else $0 = a_1\kappa_1 + a_2\kappa_2 + \cdots + a_{r-1}\kappa_{r-1} - a_r$, contradicting the independence of the a_i with respect to $\mathfrak{t}$. If $q(x)$ maps $p(a_r)$ into 1, the function $\chi(x) = q(p(x))$ is the solution.

Theorem 7.1B. If A is a simple ring with center $\mathfrak{t}$, and B is any ring with unit element, containing $\mathfrak{t}$ in its center, then every two-sided ideal of $A \times_{\mathfrak{t}} B$ is of the form $A \times_{\mathfrak{t}} \mathfrak{b}$, where $\mathfrak{b}$ is a two-sided ideal of B.

Proof. Let $\mathfrak{a}$ be any ideal of $A \times_{\mathfrak{t}} B$, and $\mathfrak{b}$ the intersection of $\mathfrak{a}$ with B. It is clear that $A \times_{\mathfrak{t}} \mathfrak{b}$ is part of $\mathfrak{a}$. Let, conversely, $\xi = \Sigma a_\nu b_\nu$ be an element of $\mathfrak{a}$. We can assume that all the a_ν are independent over $\mathfrak{t}$, since we could otherwise simplify the expression for ξ. If $\chi_i(x)$ is an analytic linear function, with coefficients in A, mapping a_i into 1 and all the other a's into 0, we find that $\chi_i(\xi) = \Sigma\chi_i(a_\nu b_\nu) = \Sigma\chi_i(a_\nu)b_\nu = b_i$, and hence $\mathfrak{a} = A \times_{\mathfrak{t}} \mathfrak{b}$.

Corollary 7.1C. If A and B are simple, and the center $\mathfrak{t}$ of A is contained in the center of B, and B is of finite degree over $\mathfrak{t}$, then $A \times_{\mathfrak{t}} B$ is simple.

For $A \times_{\mathfrak{t}} B$ is without two-sided ideals by Theorem 7.1B and has minimum condition by Theorem 6.8A.

Theorem 7.1D. Let A and B be two subrings of any ring R, with the following properties:

A and B both have the same unit element (which need not be unit element for R).

A is simple, and its center $\mathfrak{t}$ is included in the center of B.

Every element of B is commutative with every element of A.

Then the subring $A \cdot B$ of R has the structure $A \times_{\mathfrak{t}} B$. That means that there are no connecting relations between A and B other than the most trivial ones.

Proof. We construct two new rings $\bar{A}$ and $\bar{B}$ isomorphic to A and B; and having only the image $\bar{\mathfrak{k}}$ of $\mathfrak{k}$ as common elements. We form the product $\bar{A} \times_{\mathfrak{k}} \bar{B}$ and contend that the following mapping onto AB is an isomorphism. Let $\Sigma \bar{a}_\nu \bar{b}_\nu$ correspond to $\Sigma a_\nu b_\nu$. We have, first, to prove that the mapping is uniquely determined; but if $\Sigma \bar{a}_\nu \bar{b}_\nu = 0$, there is a substitution in the $\bar{b}_i$ that makes this obvious. The corresponding substitution in $\Sigma a_\nu b_\nu$ will show that this element is also 0, because this computation involves only the distributive and associative laws. That the mapping is a homomorphism under addition and multiplication is also obvious. It remains to be seen that it is a 1-1 correspondence. The set $\mathfrak{a}$ of all elements of $\bar{A} \times_{\mathfrak{k}} \bar{B}$ that are mapped onto 0 forms a two-sided ideal. According to our theorem, this ideal has the structure $\bar{A} \times_{\mathfrak{k}} \bar{\mathfrak{b}}$, where $\bar{\mathfrak{b}}$ consists of those elements $\bar{b}$ of $\bar{B}$ which are mapped onto 0. Consequently $\bar{\mathfrak{b}} = 0$, and therefore $\mathfrak{a} = 0$.

This allows us to determine the structure of the set of all analytic linear functions of R, from another point of view. These functions themselves form a ring (with iteration used as multiplication). This ring, L, contains two subrings: First, the functions $\chi_a(x) = ax$. The set of all these functions is clearly isomorphic to R as soon as R contains a unit element, and without misunderstanding we can identify this subring with R itself. Second, the functions $\chi_a^*(x) = xa$. The set of all these functions is not isomorphic to R, because

$$\chi_a^*(\chi_b^*(x)) = xba = \chi_{ba}^*(x).$$

But the mapping of R onto the χ_a^* is an anti-isomorphism, interchanging the order in a product. We identify this ring with the R* of §5.6. χ_a and χ_b^* are commutative, since $\chi_a(\chi_b^*(x)) = axb = \chi_b^*(\chi_a(x))$. If, finally, $\varkappa$ is in the center of R, $\chi_\varkappa(x) = \chi_\varkappa^*(x)$. This shows that the two rings satisfy all conditions of the preceding theorem, and therefore $R \times_{\mathfrak{k}} R^*$ is contained in L. Since, on the other hand, every element of L can be obtained from the special functions R and R* by iteration and addition, we have $L = R \times_{\mathfrak{k}} R^*$.

Now, we can obtain a partial converse to Theorem 6.10B.

Theorem 7.1E. If R is a simple ring with center $\mathfrak{k}$, and $L = R \times_{\mathfrak{k}} R^*$, then L has the minimum condition for left ideals if and only if R is of finite degree over $\mathfrak{k}$.

Proof. The "if" part follows from Theorem 6.10B, so that there remains just to prove the "only if."

Suppose, therefore, that R is not of finite degree over $\mathfrak{k}$, and $w_1, w_2, w_3, \cdots$ is an infinite sequence of elements independent over $\mathfrak{k}$. Let $\mathfrak{l}_i$ stand for the set of all analytic linear functions $\chi_i(x)$ that map $w_1, w_2, \cdots, w_i$ onto 0. Clearly $\mathfrak{l}_i \supset \mathfrak{l}_{i+1}$, and differs from it, because $\mathfrak{l}_i$ contains a linear function that maps $w_1, \cdots, w_i$ onto 0, and w_{i+1} onto 1. Since these are left ideals, we have proved the theorem.

If R is of finite dimension over $\mathfrak{k}$, the structure of L is completely determined by Theorem 7.1A.

Theorem 7.1F. If R is simple, and of finite degree over its center $\mathfrak{k}$, then $L = R \times_{\mathfrak{k}} R^*$ consists of all linear transformations of the space R, with respect to $\mathfrak{k}$.

Proof. Since every analytic linear transformation satisfies the conditions $\chi(x + y) = \chi(x) + \chi(y)$, and $\chi(\kappa x) = \kappa(\chi(x))$, for every element κ of the center, L is part of the set of all linear transformations. If, on the other hand, $w_1, w_2, \cdots, w_n$ are a $\mathfrak{k}$-basis for R, we can find a function $\chi(x)$, mapping $w_1, w_2, \cdots, w_n$ on arbitrarily selected images. But any linear transformation with respect to $\mathfrak{k}$ is uniquely determined by the images of the w_i.

Another way to state the previous theorem is this:

If R is simple and of finite degree n over $\mathfrak{k}$, the ring $R \times_{\mathfrak{k}} R^*$ is isomorphic to the ring of all $n \times n$ $\mathfrak{k}$-matrices.

2. AUTOMORPHISMS OF SIMPLE RINGS

In order to apply our representation of $R \times_{\mathfrak{k}} R^*$ as a ring of linear transformations of R, to the study of automorphisms of R, we need a theorem on spaces over simple rings. Recall Theorem 5.3E; it follows from this theorem that, if two spaces with minimum condition over a simple ring R split into the same number of irreducible components, they are isomorphic. The isomorphism consists, indeed, in mapping each irreducible component of V onto one of the irreducible components of V', and then extending that mapping to a mapping of V on V' by addition. If we have a slightly more general situation, namely, two rings R and R', that are isomorphic under the mapping $r \leftrightarrow r'$, and an R-space V and R'-space V', both of which split into the same number of irreducible components, we can do something similar. We first consider V' as an R-space, by defining the multiplication of V' by an element r of R to have the same effect as the multiplication by the image r' of r in R'. Hence there is a function σ mapping V onto V' in such a manner that for any vector v and any element r of R we have $\sigma(rv) = r\sigma(v)$. The multiplication of v by r is the original one; since $\sigma(v)$ is in V', the multiplication by r means really the multiplication by r'. We consequently have the following theorem.

Theorem 7.2A. *If R and R' are two simple isomorphic rings and V and V' are spaces over R and R', respectively, splitting into the same number of irreducible components, we can find a mapping function $\sigma(v)$ of V onto V', which is a one to one correspondence and satisfies the identity $\sigma(rv) = r'\sigma(v)$ (where r' is the image of r under our isomorphism).*

Corollary 7.2B. *Under the same conditions as before, except for the equality of the number of irreducible components of the two spaces with minimum condition, it is possible to map one of the spaces, say V, onto a part of the other by a one to one correspondence $\sigma(v)$ such that the same equation holds.*

Proof. Should V' have more components than V, we can find a subspace of V' having the same number.

Now return to the ring $L = R \times_{\mathfrak{k}} R^*$. If R is a ring with minimum condition, and we want to study a subring of L that also satisfies the minimum condition, it seems natural to restrict one of the two factors of L, by replacing it by a simple subring A of finite dimension over $\mathfrak{k}$. Such a ring A is called a simple algebra. We shall, therefore, study the subring $S = R \times_{\mathfrak{k}} A$, which consists of the linear functions

$$\chi(x) = \Sigma r_\nu x a_\nu, \tag{2.1}$$

with $r_\nu \in R$ and $a_\nu \in A$. We might now consider R as a left S-space, in the sense that χ is applied onto x, as left operator, by computing $\chi(x)$.

Since S contains the linear functions denoted by R, every S-space is also a left R-space; since R satisfies the minimal condition over R, the same is true when R is considered as a space over S. Assume now that we have in R a second algebra A', isomorphic to A, under an isomorphism $a \leftrightarrow a'$ that leaves every element of $\mathfrak{k}$ fixed. The set $S' = R \times_{\mathfrak{k}} A'$ is then isomorphic to S under an isomorphism that maps the function (2.1) onto $\chi'(x) = \Sigma r_\nu x a'_\nu$. We remark that the special function $\chi(x) = xa$ is mapped onto the function xa'. Let us now make use of Corollary

7.2B. R appears as a space over the two isomorphic rings S and S', and we may therefore assume that a function $\sigma(x)$ can be found which maps, by a 1-1 correspondence, R onto some part of itself, and satisfies for every $\chi \in S$ the equation

$$\sigma(\chi(x)) = \chi'(\sigma(x)). \tag{2.2}$$

We first specialize to $\chi(x) = rx$ and get $\sigma(rx) = r\sigma(x)$. Putting $x = 1$, and writing c for $\sigma(1)$, we have $\sigma(r) = rc$. Since r can be any element of R, we have $\sigma(x) = xc$. This mapping is a 1-1 correspondence between R and the images, so that $xc = 0$ implies $x = 0$. Since c is not a right divisor of 0, it has an inverse (the left ideals Rc form a descending chain; hence, for some n, $Rc^{n+1} = Rc^n$. In the set on the right we have c^n, and, therefore, for some x, $sc^{n+1} = c^n$; since c is not a divisor of 0, $xc = 1$). Specializing now in (2.2) to $\chi(x) = xa$, we get $xac = xca'$. Putting $x = 1$, we find

$$a' = c^{-1}ac.$$

This gives us the following theorem.

Theorem 7.2C. *Let R be a simple ring and A and A' two simple subalgebras of R containing the center $\mathfrak{k}$ of R. Any isomorphism of A to A' that leaves every element of the center $\mathfrak{k}$ fixed can be extended to an inner automorphism of R.*

Corollary 7.2D. *If R is an algebra, every automorphism of R that leaves the center fixed is inner.*

Proof. Select $A = A' = R$.

We close the section with an application. Let R be a sfield with center $\mathfrak{k}$. Consider a polynomial $f(t)$ in $\mathfrak{k}$. We are interested in zeros a of $f(t)$ which lie in R. Suppose that $f(t)$ can be factored in $\mathfrak{k}$, say, $f(t) = f_1(t) f_2(t)$. Then, since $\mathfrak{k}$ and any element a of R generate a field $\mathfrak{k}(a)$, $f(a) = f_1(a)f_2(a)$ for all $a \in R$. Thus we may, without loss of generality, limit our investigation to the study of zeros in R of polynomials $f(t)$ irreducible in $\mathfrak{k}$.

If a and b are roots in R of $f(t) = 0$, then the fields $\mathfrak{k}(a)$ and $\mathfrak{k}(b)$ are isomorphic under the correspondence $a \leftrightarrow b$, elements of $\mathfrak{k}$ fixed. Now apply Theorem 7.2C with $A = \mathfrak{k}(a)$ and $A' = \mathfrak{k}(b)$, and we obtain an element c of R such that $b = cac^{-1}$. Conversely, if a is a zero of $f(t)$ and $c \in R$, then cac^{-1} is again a zero of $f(t)$. The elements a and cac^{-1} are called *conjugates* in R. We have proved the following theorem.

Theorem 7.2E. *Let* R *be a sfield with center* $\mathfrak{k}$, *and let* $f(t)$ *be an irreducible* $\mathfrak{k}$*-polynomial with the zero* a *in* R. *Then* b *is a zero of* $f(t)$ *if and only if* a *and* b *are conjugate in* R.

For instance, if R is the sfield of real quaternions and $f(t) = t^2 + 1$, then i and j are zeros of $f(t)$, and so there is a quaternion c such that $j = cic^{-1}$.

3. COMMUTATORS OF SUBRINGS

In order to study the way in which a simple subring is embedded in a larger ring, we need to define the degree of a ring, not only over a subfield, but over a simple subring. To do this we commence with the following lemma.

Lemma 7.3A. *Let* S *be a simple subring of any ring* R, *and let* R *and* S *have the same unit element. If* $\mathfrak{l}_1$ *and* $\mathfrak{l}_2$ *are two minimal ideals of* S, *then any isomorphism of* $\mathfrak{l}_1$ *onto* $\mathfrak{l}_2$, *considered as* S*-spaces, can be extended to an isomorphism of* $R\,\mathfrak{l}_1$ *to* $R\,\mathfrak{l}_2$, *considered as* R*-spaces.*

Proof. If e_1 and e_2 are idempotents of $\mathfrak{l}_1$ and $\mathfrak{l}_2$, let c_2 be the image of e_1 under the mapping of $\mathfrak{l}_1$ onto $\mathfrak{l}_2$. Then $x = xe_1$ of $\mathfrak{l}_1$ is mapped onto xc_2 of $\mathfrak{l}_2$. Therefore if we multiply the set $Re_1 = RSe_1 = R\,\mathfrak{l}_1$ from the right by c_2, this mapping is an R-homomorphism and an extension of our mapping of $\mathfrak{l}_1$ onto $\mathfrak{l}_2$. The set of images is $Rc_2 = RSc_2 = R\,\mathfrak{l}_2$.

Let, now, c_1 be the image of e_2 under the inverse of our initial isomorphism, so that $x \in \mathfrak{l}_2$ is hereby mapped onto xc_1. Then $c_2c_1 = e_1$. If, now, $xe_1c_2 = 0$, we would have $xe_1c_2c_1 = xe_1 = 0$. Our mapping is therefore an isomorphism.

Now let $S = \mathfrak{l}_1 + \mathfrak{l}_2 + \cdots + \mathfrak{l}_r$ be a splitting in minimal left ideals and $1 = \Sigma e_\nu$ be the corresponding splitting in orthogonal idempotents. We easily deduce $R = Re_1 + Re_2 + \cdots + Re_r$. This is a direct sum, since $x_1e_1 + x_2e_2 + \cdots + x_re_r = 0$ shows after right multiplication by e_i that $x_ie_i = 0$. We now consider R as a left S-space. Each of the components Re_i is then also a left S-space, and they are isomorphic, since they are even isomorphic if considered as R-spaces. Therefore we can break each of them into an equal number of irreducible S-spaces. This will lead, after regrouping, to a splitting of R into a finite or an infinite sum of left S-spaces, each of which is a direct sum of precisely r components, and is thus isomorphic to S itself. An S-subspace of R that is isomorphic to S has the following structure: let w be the image of 1 if we map S onto the subspace. x of S is then mapped onto xw of the subspace, and since the mapping is an isomorphism, xw = 0 with x in S implies x = 0. On doing this with each of our spaces, we see that there exist elements w_i such that $R = \Sigma S w_i$. Any finite number of w_i are independent with respect to S, in the sense that a linear combination of them with coefficients in S is 0 if and only if all coefficients (not merely all terms) of the sum are 0. The number of irreducible S-components of the S-space R is r times as large as the number of w_i. Therefore the number of w_i is invariant; we shall call it the left degree of R over S, and will denote it by (R : S). Now, in precisely the same way as in case of fields, we can prove that $(R : S_1) = (R : S)(S : S_1)$, if S_1 is a simple subring of S.

If R is a ring and S a subset of R, we denote by R^S the set of all elements of R which are commutative with every element of S.

Lemma 7.3B. If R and T are rings with unit element whose centers contain the field $\mathfrak{k}$, and S is a subset of R, then $(R \times_{\mathfrak{k}} T)^S = R^S \times_{\mathfrak{k}} T$. If $S \subseteq R$ and $U \subseteq T$, and S and U are rings containing $\mathfrak{k}$, then $(R \times_{\mathfrak{k}} T)^{S \times_{\mathfrak{k}} U} = R^S \times_{\mathfrak{k}} T^U$.

Proof. Clearly $(R \times_{\mathfrak{k}} T)^S \supseteq R^S \times_{\mathfrak{k}} T$. Let $\Sigma r_\nu t_\nu$ be any element of the left side, with linearly independent t_ν. If s is any element of S, then $\Sigma(s r_\nu - r_\nu s)t_\nu = 0$. Since the t_ν are linearly independent, we see (by Lemma 6.4C) that every r_ν is in R^S. R^R is, of course, the center of R.

Let R be a ring and T be an algebra, both with center $\mathfrak{k}$. Then $(R \times_{\mathfrak{k}} T)^{R \times_{\mathfrak{k}} T} = \mathfrak{k} \times_{\mathfrak{k}} \mathfrak{k} = \mathfrak{k}$, so that $\mathfrak{k}$ is also the center of $R \times_{\mathfrak{k}} T$. Assume now that R and T are simple rings, that each contains a simple algebra containing $\mathfrak{k}$, and that these two algebras are isomorphic under a correspondence leaving the elements of $\mathfrak{k}$ invariant. We shall denote both these algebras, though different, by the same letter, A, since it will be clear from the formulas which is meant. The simple ring $R \times_{\mathfrak{k}} T$ now contains the algebra A in two ways; consequently there is an inner automorphism of $R \times_{\mathfrak{k}} T$ that will map one onto the other. Computing $(R \times_{\mathfrak{k}} T)^A$ for both algebras A, we get the sets $R^A \times_{\mathfrak{k}} T$, and $R \times_{\mathfrak{k}} T^A$. These are, therefore, isomorphic under the same inner automorphism. Computing further, we get

$$R^{(R^A)} = (R \times_{\mathfrak{k}} T)^{(R^A \times_{\mathfrak{k}} T)}$$

is isomorphic to

$$T^{(T^A)} = (R \times_{\mathfrak{k}} T)^{(R \times_{\mathfrak{k}} T^A)}$$

under the same isomorphism.

Should we be able to prove that R^A is simple, we could compute the degree of $R \times_{\mathfrak{k}} T$ over $R^A \times_{\mathfrak{k}} T$; since an inner automorphism of $R \times_{\mathfrak{k}} T$ maps $R^A \times_{\mathfrak{k}} T$ onto $R \times_{\mathfrak{k}} T^A$, the degree of $R \times_{\mathfrak{k}} T$ over $R^A \times_{\mathfrak{k}} T$ should be the same as the degree of $R \times_{\mathfrak{k}} T$ over $R \times_{\mathfrak{k}} T^A$. The reader can easily deduce the equality $(R : R^A) = (T : T^A)$. The proof for the simplicity will be given soon.

If we know only that our algebra A is contained in R, we have to construct a T. Assume that the degree $(A : \mathfrak{k})$ is n. Consider A as a $\mathfrak{k}$-space, and denote by M_n the set of all linear transformations of A over $\mathfrak{k}$, that is, the algebra of $n \times n$ matrices over $\mathfrak{k}$. The degree $(M_n : \mathfrak{k})$ will be n^2. M_n contains a subalgebra isomorphic to A, namely, the set of all linear transformations of A that consist of multiplying A from the left by an element of A. Calling this ring of linear transformations also A, we now ask for M_n^A, that is, for the linear transformations σ that satisfy the condition $\sigma(ax) = a\sigma(x)$, for all a and x in A. Putting $x = 1$, denoting $\sigma(1)$ by c and writing x instead of a, we see $\sigma(x) = xc$. It follows easily that the set of these linear transformations is isomorphic to A^*, so that we have $M_n^A = A^*$. (We did not assume that the center of A is $\mathfrak{k}$; the center might be larger, so that M_n would possibly not be equal to $A \times_{\mathfrak{k}} A^*$.)

Substituting M_n for T in our previous result, we can prove

Theorem 7.3C. If R is simple and A is a simple subalgebra of R containing the center $\mathfrak{k}$ of R, and if, finally, $(A : \mathfrak{k}) = n$, we have:

(1). $R^A \times_{\mathfrak{k}} M_n$ and $R \times_{\mathfrak{k}} A^*$ are isomorphic under an inner automorphism of $R \times_{\mathfrak{k}} M_n$.

(2). R^A is simple.

(3). $R^{(R^A)} = A$.

(4). $(R : R^A) = n$.

Proof. The first statement is clear. Since A^* is simple, $R \times_{\mathfrak{k}} A^*$ is simple. This proves that $R^A \times_{\mathfrak{k}} M_n$ and, consequently, R^A is simple. As for the third statement, $R^{(R^A)}$ is mapped by our inner automorphism onto $M_n^{(M^A)} = M^{A*}$, which is the A contained in M. Since we know that our isomorphism maps precisely the A of R onto the A of M_n, we get our statement. The fourth statement follows from the fact that $(M_n : A^*) = n$. (For $(R : R^A) = (R \times_{\mathfrak{t}} M : R^A \times_{\mathfrak{t}} M) = (R \times_{\mathfrak{t}} M : R \times_{\mathfrak{t}} A^*) = (M : A^*))$. This follows in turn from $(M_n : \mathfrak{t}) = n^2$ and $(A^* : \mathfrak{t}) = n$.

There is one very general theorem (Theorem 7.3F) concerning simple subalgebras A of rings, that holds for any ring R whose center contains the center $\mathfrak{t}$ of A and whose unit element is contained in $\mathfrak{t}$. The proof of this theorem will depend on the following lemma.

Lemma 7.3D. *Assume that an analytic linear function* $\chi(x) = \Sigma a_\nu x a'_\nu$, *with coefficients in* A, *is* 0 *for all* $x \in A$. *Then* $\chi(x)$ *is* 0 *for all* $x \in R$.

Proof. Since we can express the a'_ν by linearly independent elements of A(that is, independent over $\mathfrak{t}$) and collect terms, we can assume from the beginning that the a'_ν are linearly independent. The necessary operation for collecting is the interchange of x with an element of the center $\mathfrak{t}$ of A; this can be done, according to our assumption, for all $x \in R$.

We contend that now all the a_ν are 0. To show this, we consider $\chi(x)$ for values of A only. The linear functions of A form a ring isomorphic to $A \times_{\mathfrak{t}} A^*$, and under this isomorphism our linear function will correspond to $\Sigma a_\nu a'^*_\nu$. Since $\chi(x) = 0$ for all $x \in A$, we have $\Sigma a_\nu a'^*_\nu = 0$, and the linear independence of the a'_ν shows that all the a_ν are 0. But this means that $\chi(x) = 0$ for every x in R.

Our first aim is to compute R^A; we contend that this can be done in the following way.

Theorem 7.3E. *Let* $\chi(x)$ *be any analytic linear function with coefficients in* A *which is not identically zero and which maps every element of* A *onto an element of* $\mathfrak{t}$. *The values* $\chi(x)$, *for all* x *in* R, *will then range precisely over* R^A.

Proof. If a is any element of A, $\chi(x)a - a\chi(x)$ is a linear function with coefficients in A that is 0 for every $x \in A$. According to our lemma, it is 0 for every x in R. Since a is arbitrary, this proves that $\chi(x) \in R^A$.

To prove that we get every element of R^A, take any element a such that $\chi(a) = \kappa \neq 0$. Since we can replace a by $\kappa^{-1}a$, we can assume that $\chi(a) = 1$. If z is any element of R^A, we get $\chi(za) = z\,\chi(a) = z$, because we can interchange z with every coefficient.

Theorem 7.3F. If A is a simple subalgebra of a ring R whose center $\mathfrak{k}$ of A and whose unit element is contained in $\mathfrak{k}$, then $R = R^A \times_{\mathfrak{k}} A$.

Proof. Let $a_1, a_2, \cdots, a_n$ be a $\mathfrak{k}$-basis for A. Let $\chi_i(x)$ be the analytic linear function with coefficients in A that takes a_i into 1 and all the other a's into 0. The linear function $\chi(x) = a_1\chi_1(x) + a_2\chi_2(x) + \cdots + a_n\chi_n(x) - x$ is 0 for $x = a_1, a_2, \cdots, a_n$, and therefore for every x in A. This proves that for every x in R we have $x = a_1\chi_1(x) + a_2\chi_2(x) + \cdots + a_n\chi_n(x)$. Since the values of $\chi_i(x)$ range through R^A, we see that $R = A \cdot R^A$. That the dot can be replaced by the cross follows from Theorem 7.1D.

Corollary 7.3G. If A is a simple algebra with unit element e and center $\mathfrak{k}$, contained in a ring R (for which we make no assumption), then $eR^{\mathfrak{k}} = A \times_{\mathfrak{k}} eR^A$.

Proof. Consider first the ring $(eRe)^{\mathfrak{k}}$. Obviously the center of this ring contains $\mathfrak{k}$. e is unit element for this ring, and the previous theorem can be applied. But $((eRe)^{\mathfrak{k}})^A = (eRe)^A = eR^A$.

Theorem 7.3H. Let R be simple and let $\mathfrak{k}$ be its center and T a subring containing $\mathfrak{k}$. If $R = w_1T + w_2T + \cdots + w_nT$, then R^T is an algebra A, and $(A : \mathfrak{k}) \leqq n$.

Furthermore, these two statements are equivalent:

(1). T is the commutator of a simple subalgebra of R.

(2). T is simple, (R : T) is finite, and all homomorphisms of R into itself, considering R as right T-space, are analytic.

Proof. Let T be any subring of R such that $R = w_1T + w_2T + \cdots + w_nT$. Consider the set S of all homomorphisms of R regarded as a right T-space. This ring S contains as subring the analytic transformations among them. Let $\sigma(x) = \Sigma r_\nu x r'_\nu$, and assume the r_ν linearly independent with respect to $\mathfrak{k}$. Since, for every $t \in T$, $\sigma(xt) - \sigma(x)t = 0$, we get the equations $tr'_\nu - r'_\nu t = 0$, so that all r'_ν belong to $A = R^T$. This subring is, therefore, to be denoted by $R \times_{\mathfrak{k}} A^*$. It contains as further subring the left multiplications of x by elements of R, and this ring of transformations we have always denoted by R.

Since the degree $(A : \mathfrak{k}) = (R \times_{\mathfrak{k}} A^* : R)$, our inequality will be proved if we estimate the degree $(S : R)$. Every homomorphism $\sigma(x)$ of S will map $w_1, w_2, \cdots, w_n$ onto certain elements $a_1, a_2, \cdots, a_n$ of R. If these elements are known, and $x = w_1t_1 + w_2t_2 + \cdots + w_nt_n$, then its image will be $a_1t_1 + a_2t_2 + \cdots + a_nt_n$. Let us denote $\sigma(x)$ by the n-dimensional vector $(a_1, a_2, \cdots, a_n)$. We are interested in the behavior of S considered as left R-space, where R here is the set of transformations $\rho(x) = rx$. Iterating our previous σ with that transformation would lead to $(ra_1, ra_2, \cdots, ra_n)$. This shows that S is isomorphic to a certain part of the set of all n-dimensional vectors over R. The statement concerning the degrees can also be interpreted as a statement about the number of irreducible components; the degree has only to be multiplied by the number of irreducible components R has. This makes it clear that $(S : R) \leqq n$. (From previous work we know that this degree exists.) Hence $(A : \mathfrak{k}) \leqq n$. It is, furthermore, clear that $(S : R) = n$ if and only if S fills out the whole space of n-vectors. This means that there are linear transformations mapping the w_i onto arbitrary images. But then the w_i have to be linearly independent, because $w_1t_1 + w_2t_2 + \cdots + w_nt_n = 0$ would yield $t_i = 0$ if we applied the homomorphism mapping w_i onto 1 and the others onto 0. If, conversely, the w_i are linearly independent, and we map $w_1t_1 + w_2t_2 + \cdots + w_nt_n$ onto $a_1t_1 + a_2t_2 + \cdots + a_nt_n$, this clearly gives an element of S.

If A is a simple subalgebra of R, and T is taken as R^A, we may apply Theorem 7.3C. So the w_i can be taken as linearly independent over T and $n = (R : R^A) = (A : \mathfrak{k}) = (S : R)$. Since $(R \times_{\mathfrak{k}} A^* : R) = n$ and $R \times_{\mathfrak{k}} A^* \subsetneqq S$, we get $R \times_{\mathfrak{k}} A^* = S$, which means that all our homomorphisms are analytic.

Conversely, if T is simple and $(R : T)$ is finite, we can take the w's linearly independent. If we assume that all our homomorphisms are analytic, so that $S = R \times_{\mathfrak{k}} A^*$, we get first $n = (R \times_{\mathfrak{k}} A^* : R) = (A : \mathfrak{k})$. We next get that S is simple, because it consists of all homomorphisms of the n-dimensional T-space R, and is, therefore, isomorphic to the set of all $n \times n$ matrices with elements in T, or $T \times_{\mathfrak{k}} M_n$. So $R \times_{\mathfrak{k}} A^*$ is simple, which implies that A is simple. Since R^A must be of degree n under R, and certainly contains T, we find that $T = R^A$, which finishes the proof.

Corollary 7.3I. *If A is a simple subalgebra of R containing the center $\mathfrak{t}$ of R (R simple), and $\sigma(x)$ is a homomorphism of R satisfying the equations $\sigma(xt) = \sigma(x)t$ and $\sigma(tx) = t\sigma(x)$, for all elements t of $T = R^A$, $\sigma(x)$ can be written $\Sigma a_\nu x a'_\nu$, where the a_ν as well as the a'_ν are in A.*

This is proved by the method used in the first part of the proof of the preceding theorem.

Theorem 7.3J. *Under the same assumptions as in the preceding corollary every automorphism of R that leaves every element of $T = R^A$ fixed consists of a transformation by an element of A.*

Proof. Since $\sigma(xt) = \sigma(x)\sigma(t) = \sigma(x)t$ and $\sigma(tx) = t\sigma(x)$, our corollary gives $\sigma(x) = \Sigma a_\nu x a'_\nu$. We can assume that the a_ν and the a'_ν are linearly independent with respect to $\mathfrak{t}$. Let $\chi(x) = \Sigma b_\mu x b'_\nu$ be an analytic linear function of A which takes a_1 into 1 and the other a_i into 0. Compute $\Sigma_\mu b_\mu \sigma(b'_\mu x)$. On the one hand, it is $\{\Sigma_\mu b_\mu \sigma(b'_\mu)\}\sigma(x)$, and therefore is of the form $a\sigma(x)$. On the other hand, it is $\Sigma_{\nu,\mu} b_\mu a_\nu b'_\mu x a'_\nu = \Sigma_\nu \chi(a_\nu) x a'_\nu = x a'_1$. Here a'_1 is not 0 because it is one of a set of linearly independent elements. We therefore get $a\sigma(x) = x a'_1$. Let $\lambda(x) = \Sigma c_\nu x c'_\nu$ be such that $\lambda(a'_1) = 1$. Compute $\Sigma\lambda a\ (xc_\nu)c'$. On the one hand, it has the form $a\sigma(x)b$; on the other hand, it is $\Sigma x c_\nu a'_1 c'_\nu = x$. Hence $a\sigma(x)b = x$. Putting $x = 1$, we see that $b = a^{-1}$, and hence $\sigma(x) = a^{-1}xa$. That a belongs to A follows from the corollary.

CHAPTER VIII

SPLITTING FIELDS AND CROSSED PRODUCTS*

1. INTRODUCTION

In this chapter we show that computation in simple rings can be reduced to computations in a commutative field. We also show how to construct simple algebras by use of commutative fields.

2. SIMILARITY OF SIMPLE ALGEBRAS

It is convenient to state the Wedderburn Theorem 5.1A in terms of direct products. To do this we need the following almost obvious lemma.

Lemma 8.2A. Let $\mathfrak{k}$ be any field, R be any ring for which $\mathfrak{k}$ is defined as a left and right operator domain in such a manner that the unit element of $\mathfrak{k}$ is unit operator. Let R_n be the ring of n x n matrices with elements in R, and M_n the ring of n x n matrices with elements in $\mathfrak{k}$. Then $R_n \cong R \times_{\mathfrak{k}} M_n$.

Proof. Let e_{ij} denote the matrix which has the unit element 1 of $\mathfrak{k}$ in the i-th row and j-th column, and has 0 elsewhere. The elements e_{ij} are not necessarily contained in R_n, but every element of R_n can be uniquely represented in the form $\Sigma\gamma_{ij}e_{ij}$, $\gamma_{ij} \in R$. But the elements of $R \times_{\mathfrak{k}} M_n$ are also uniquely represented in this way, with the same rules of addition and multiplication.

Corollary 8.2B. If R is any simple ring with center $\mathfrak{k}$, then R contains a sfield $S \supseteq \mathfrak{k}$ and a total matrix algebra M, with elements in $\mathfrak{k}$, such that $R = S \times_{\mathfrak{k}} M$. If S', M' are another pair of subrings of R satisfying the same condition as S and M, then $S' \cong S$ and $M' \cong M$, under automorphisms of R leaving the elements of $\mathfrak{k}$ fixed. $\mathfrak{k}$ is the center of S.

Proof. The corollary follows at once from 8.2A and the Wedderburn Theorem 5.1A. Since, by 7.3B, the

*Prepared by Dr. George Whaples from lectures of Professor Artin at Indiana University.

the center of $S \times_{\mathfrak{k}} M$ is the Kronecker product of the centers of S and of M, the center of S is $\mathfrak{k}$.

We will use M_n to stand for the algebra of $n \times n$ matrices over the field $\mathfrak{k}$. The n will be omitted unless absolutely necessary.

Lemma 8.2C. $M_n \times_{\mathfrak{k}} M_m = M_{nm}$.

Proof. Let $e_{ij}(i = 1 \cdots n)$ be generators of M_n over $\mathfrak{k}$, and $e'_{ij}(i', j' = 1 \cdots m)$ generators of M_m over $\mathfrak{k}$. Then $M_n \times_{\mathfrak{k}} M_m$ is generated by the elements $E_{ii',jj'} = e_{ij} \cdot e_{i'j'}$. The E's satisfy the relations

$$(2.1) \qquad E_{ii',kk'} E_{\ell\ell',jj'} = \begin{cases} E_{ii',jj'} & \text{if } kk' \text{ is the same pair as } \ell\ell'; \\ 0 & \text{if } kk' \text{ is not the same pair as } \ell\ell'. \end{cases}$$

So the $E_{ii',jj'}$ generate an $nm \times nm$ total matrix algebra; for M_{nm} is generated by elements $E_{I,J}$ where I, J run through a set of nm objects, and it is immaterial whether those objects are the rational integers usually used or are nm different pairs of integers.

Two simple rings R and T are called similar (notation: $R \sim T$) if they have the same sfield component, that is, if there is a sfield S such that $R \cong S \times_{\mathfrak{k}} M_n$, $T \cong S \times_{\mathfrak{k}} M_m$. Clearly similarity is a symmetric, reflexive, and transitive relation, and divides the set of all simple rings with center $\mathfrak{k}$ into classes of similar rings. It follows from 8.2C and the associativity of $\times_{\mathfrak{k}}$ that, if R and T have the same center $\mathfrak{k}$, $R \sim T$ if and only if there exist M_m, M_n such that $R \times_{\mathfrak{k}} M_m = T \times_{\mathfrak{k}} M_n$. From the same lemmas it follows that, if $R \sim R'$, $T \sim T'$, then $R \times_{\mathfrak{k}} T \sim R' \times_{\mathfrak{k}} T'$. (For if $R = S \times_{\mathfrak{k}} M_n$, $T = L \times_{\mathfrak{k}} M_m$, $R' = S \times_{\mathfrak{k}} M_{n'}$, $T' = L \times_{\mathfrak{k}} M_{m'}$, then $R \times_{\mathfrak{k}} T = (S \times_{\mathfrak{k}} L) \times_{\mathfrak{k}} M_{mn}$ and $R' \times_{\mathfrak{k}} T' = (S \times_{\mathfrak{k}} L) \times_{\mathfrak{k}} M_{n'm'}$.)

Theorem 8.2D. The classes of similar simple algebras with center $\mathfrak{k}$ form a commutative group under $\times_{\mathfrak{k}}$.

Proof. The class of total matrix algebras is unit element. (If M is a total matrix algebra, we write $M \sim 1$.) It is only necessary to check existence of an inverse. But if A is an algebra with center $\mathfrak{k}$, the inverse-isomorphic algebra A^* also has center $\mathfrak{k}$, and, by 7.1F, $A \times_{\mathfrak{k}} A^* \sim 1$.

Note that the set of classes of similar rings with center $\mathfrak{k}$ does not form a group--indeed, 7.1E shows that if S is a sfield which is not of finite degree over $\mathfrak{k}$,

$S \times_{\mathfrak{k}} S^*$ does not have the minimum condition. In this chapter we restrict ourselves to simple algebras. A sfield which is also an algebra is called a division algebra. Obviously, the sfield component of any simple algebra is a division algebra.

3. SPLITTING FIELDS OF SIMPLE ALGEBRAS

Let A be a simple algebra with center $\mathfrak{k}$, and $\mathfrak{K}$ a field containing $\mathfrak{k}$. By Corollary 7.1C and Lemma 7.3B, $A \times_{\mathfrak{k}} \mathfrak{K}$ is a simple algebra with center $\mathfrak{K}$, and $(A \times_{\mathfrak{k}} \mathfrak{K} : \mathfrak{K}) = (A : \mathfrak{k})$. If $A \times_{\mathfrak{k}} \mathfrak{K}$ is total matrix algebra, $\mathfrak{K}$ is called a splitting field of A. Clearly, in asking whether $\mathfrak{K}$ is a splitting field of A, we need only consider the division algebra component of A, for $(D \times_{\mathfrak{k}} M) \times_{\mathfrak{k}} \mathfrak{K} = (D \times_{\mathfrak{k}} \mathfrak{K}) \times_{\mathfrak{k}} M$.

Theorem 8.3A. Let $\mathfrak{K}$ be an algebraic extension field of $\mathfrak{k}$, D a division algebra with center $\mathfrak{k}$, and r the smallest integer such that $A = D \times_{\mathfrak{k}} M_r$ contains a subfield isomorphic to $\mathfrak{K}$ (under an isomorphism leaving elements of $\mathfrak{k}$ invariant). Then these three statements are equivalent:

(1). $\mathfrak{K}$ is a splitting field of D.

(2). $\mathfrak{K}$ is a maximal subfield of A (that is, is contained in no larger subfield of A).

(3). $A^{\mathfrak{K}} = \mathfrak{K}$.

Proof. First, we note that if r is minimal, $A^{\mathfrak{K}}$ is a sfield. For suppose $A^{\mathfrak{K}}$ were not a sfield. Then, since $A^{\mathfrak{K}}$ is simple, $A^{\mathfrak{K}}$ would contain some matrix algebra M_ν over $\mathfrak{k}$, with $\nu > 1$. Then, since every element of $\mathfrak{K}$ is commutative with M_ν, $\mathfrak{K} \subseteq A^{M_\nu}$. But by Theorem 7.3F, $A = A^{M_\nu} \times_{\mathfrak{k}} M$; this equation is possible only if $A^{M_\nu} = D \times_{\mathfrak{k}} M_s$, where $s = r/\nu$. But then $\mathfrak{K} \subseteq D \times_{\mathfrak{k}} M_s$, so that r is not minimal.

Now let r be minimal; then $A^{\mathfrak{K}}$ is a division algebra and by 7.3C, (1), $A^{\mathfrak{K}} \sim A \times_{\mathfrak{k}} \mathfrak{K} \sim D \times_{\mathfrak{k}} \mathfrak{K}$. So $A^{\mathfrak{K}}$ is the sfield component of $D \times_{\mathfrak{k}} \mathfrak{K}$. Hence $D \times_{\mathfrak{k}} \mathfrak{K}$ is a total matrix algebra over $\mathfrak{K}$ if and only if $A^{\mathfrak{K}} = \mathfrak{K}$, that is, (1) $\leftrightarrow$ (3).

To show that (2) $\leftrightarrow$ (3) we note that when r is minimal $A^{\mathfrak{K}}$ is a sfield which clearly includes $\mathfrak{K}$. If α is an element of $A^{\mathfrak{K}}$, not in $\mathfrak{K}$, then $\mathfrak{K}[\alpha]$ is a commutative algebra without divisors of 0 and is thus, according to Theorem 6.10A, a field; hence $\mathfrak{K}$ is not a maximal subfield of A. Conversely, if A contains an extension field $\mathfrak{K}'$ of $\mathfrak{K}$, clearly $\mathfrak{K}' \subseteq A^{\mathfrak{K}}$, and so $A^{\mathfrak{K}} \neq \mathfrak{K}$.

One should note that 8.3A gives a criterion for testing whether any algebraic extension $\mathfrak{K}$ of $\mathfrak{k}$ is a splitting field; for if $(\mathfrak{K} : \mathfrak{k}) = n$, $D \times_{\mathfrak{k}} M_n$ certainly contains a subfield isomorphic to $\mathfrak{K}$, so that there is some minimal $r(\leq n)$ satisfying the hypothesis of the theorem. Hence we can use (2) or (3) of the theorem to determine if $\mathfrak{K}$ is a splitting field.

Corollary 8.3B. If D is a division algebra with center $\mathfrak{k}$, then D contains a splitting field $\mathfrak{K}$. For every splitting field $\mathfrak{K}$ contained in D, $(D : \mathfrak{k}) = (\mathfrak{K} : \mathfrak{k})^2$; thus the degree of every simple algebra over its center is a perfect square.

Proof. Since the degree of any subfield of D is less than the degree of D, D contains a maximal subfield $\mathfrak{K}$; $\mathfrak{K}$ is a splitting field by 8.3A.

By 7.3C, $(D: D^{\mathfrak{K}}) = (\mathfrak{K} : \mathfrak{k})$. If $\mathfrak{K}$ is a splitting field, $D^{\mathfrak{K}} = \mathfrak{K}$; $(D : \mathfrak{K}) = (\mathfrak{K} : \mathfrak{k})$, so $(D : \mathfrak{k}) = (\mathfrak{K} : \mathfrak{k})^2$. If A is a simple algebra with center $\mathfrak{k}$, A can be written as $D \times_{\mathfrak{k}} M_n$; $(A : \mathfrak{k}) = (D : \mathfrak{k})\ (M_n : \mathfrak{k})$, and both factors are squares.

Corollary 8.3C. If $\mathfrak{K}$ is a splitting field of a division algebra D with center $\mathfrak{k}$, $(\mathfrak{K} : \mathfrak{k})$ is a multiple of $\sqrt{(D : \mathfrak{k})}$.

For $\mathfrak{K}$ is a maximal subfield of an algebra $A = D \times_{\mathfrak{k}} M_r$, so that $(\mathfrak{K} : \mathfrak{k})^2 = (D : \mathfrak{k})r^2$.

For application to crossed products we need

Theorem 8.3D. Every division algebra with center $\mathfrak{k}$ contains a splitting field which is separable over $\mathfrak{k}$.

Proof. $\mathfrak{K}$ is called separable over $\mathfrak{k}$ when every element of $\mathfrak{K}$ satisfies an equation which has coefficients in $\mathfrak{k}$ and does not have multiple roots in any extension field of $\mathfrak{k}$. An element of $\mathfrak{K}$ which satisfies such an equation is said to be separable over $\mathfrak{k}$. As part of the Galois theory, which must be taken for granted in this chapter, we use these facts (for proofs see van der Waerden, Moderne Algebra, Vol. I, Chapter 5, or any modern text on Galois theory).

(1) If the field $\mathfrak{k}(\xi)$ is an algebraic extension of $\mathfrak{k}$ which is larger than $\mathfrak{k}$, but which contains no separable extension of $\mathfrak{k}$ (other than $\mathfrak{k}$ itself), then $\mathfrak{k}$ is of characteristic p (for some prime p) and ξ satisfies an equation of form $\xi^{p^\nu} = \alpha \varepsilon \mathfrak{k}$.

(2) If $\mathfrak{k}'$ is separable over $\mathfrak{k}$ and $\mathfrak{K}$ separable over $\mathfrak{k}'$, then $\mathfrak{K}$ is separable over $\mathfrak{k}$.

To prove 8.3D we first show: _If D is any division algebra ($\neq \mathfrak{k}$) with center $\mathfrak{k}$, D contains a subfield which is separable over $\mathfrak{k}$ and greater than $\mathfrak{k}$._ For suppose that every element d of D is inseparable over $\mathfrak{k}$; then every $d \in D$ satisfies an equation $d^{p^\nu} = \alpha \in \mathfrak{k}$. Let $d_1, d_2, \cdots d_n$ be a basis of D with respect to $\mathfrak{k}$ and consider the "general element"

$$d = x_1 d_1 + x_2 d_2 + \cdots + x_n d_n,$$

where the $x_1, x_2, \cdots, x_n$ are variables. Let μ be an integer so large that every special element f of D satisfies an equation $f^{p^\mu} = \alpha \in \mathfrak{k}$. Such a μ exists since the degree of the irreducible equation for f is bounded by the degree of D. Computing d^{p^μ} and collecting terms, we get

$$d^{p^\mu} = X_1 d_1 + X_2 d_2 + \cdots + X_n d_n,$$

where the $X_1, \cdots, X_n$ are polynomials in the x_i with coefficients in $\mathfrak{k}$. We may assume $d_1 = 1$. For any special values of the x_i the right side is in $\mathfrak{k}$, so that $X_2 = X_3 = \cdots = X_n = 0$ for all special values of the x_i. Now $\mathfrak{k}$ may be assumed to be infinite, since a finite field has no inseparable extensions; it follows that the polynomials $X_2, X_3, \cdots, X_n$ are identically zero.

Let Ω be a splitting field of D. The above argument shows that also in $D \times_{\mathfrak{k}} \Omega$ every element ξ will satisfy an equation $\xi^{p^\mu} = \omega \in \Omega$. But this is absurd, since $D \times_{\mathfrak{k}} \Omega$ is a total matrix algebra, and no power of the matrix unit e_{11} is ever in Ω, since $e_{11}^m = e_{11}$ for all m.

Now let D be our original division algebra with center $\mathfrak{k}$, and suppose that $\mathfrak{K}$ is a maximal _separable_ subfield of D (that is, that $\mathfrak{K}$ is contained in no larger separable subfield). Suppose that $\mathfrak{K}$ is not a splitting field; then, by 8.3A, $D^{\mathfrak{K}}$ is a division algebra with center $\mathfrak{K}$ and bigger than $\mathfrak{K}$. Hence $D^{\mathfrak{K}}$ contains a field $\mathfrak{K}'$ which is separable over $\mathfrak{K}$ and bigger than $\mathfrak{K}$. This is a contradiction, for then $\mathfrak{K}' \subseteq D$ and is separable over $\mathfrak{k}$, whereas $\mathfrak{K}$ was a maximal subfield of D with this property.

Theorem 8.3E. If A is any simple algebra with center $\mathfrak{k}$, A has a splitting field which is normal separable over $\mathfrak{k}$.

This follows easily from

Lemma 8.3F. If A is any ring which is a $\mathfrak{k}$-space, and Ω and $\mathfrak{K}$ are fields such that $\Omega \supseteq \mathfrak{K} \supseteq \mathfrak{k}$, then $A \times_{\mathfrak{k}} \Omega \cong (A \times_{\mathfrak{k}} \mathfrak{K}) \times_{\mathfrak{K}} \Omega$.

Proof of Lemma. First consider the rings merely as additive groups. It is clear that, if we are given a finite set of elements of Ω, we can find a finite set of elements $\varkappa_\nu \omega_\mu$, the $\omega_\mu \in \Omega$ independent over $\mathfrak{K}$, the $\mathfrak{K}_\nu \in \mathfrak{K}$ independent over $\mathfrak{k}$, in terms of which each of our given elements can be expressed as a $\mathfrak{k}$-linear combination; hence we can write any element of $A \times_{\mathfrak{k}} \Omega$ as $x = \Sigma_{\nu,\mu} a_{\nu\mu}(\varkappa_\nu \omega_\mu)$; and if the $\varkappa_\nu$ and ω_μ are independent over $\mathfrak{k}$ and $\mathfrak{K}$ respectively, the elements $\varkappa_\nu \omega_\mu$ are independent over $\mathfrak{k}$, hence $x = 0$ only when each $a_{\nu\mu} = 0$. But this shows that the element x of $A \times_{\mathfrak{k}} \Omega$ is 0 if and only if the corresponding element $\Sigma_\mu(\Sigma_\nu a_{\nu\mu} \varkappa_\nu)\omega_\mu$ of $(A \times_{\mathfrak{k}} \mathfrak{K}) \times_{\mathfrak{K}} \Omega$ is 0; hence the additive groups of our two rings are isomorphic. Then the rings are isomorphic, since multiplication is defined by the distributive law and by multiplication in the rings A and Ω.

Proof of Theorem 8.3E. Let D be the division algebra component of A and let $\mathfrak{K}$ be any splitting field of D which is separable over $\mathfrak{k}$; such fields exist by Theorem 8.3D. Let Ω be any separable extension of $\mathfrak{K}$ which is normal over $\mathfrak{k}$; then, by the lemma, $D \times_{\mathfrak{k}} \Omega \cong (D \times_{\mathfrak{k}} \mathfrak{K}) \times_{\mathfrak{K}} \Omega$, so that Ω is a splitting field.

4. CROSSED PRODUCTS

Let B be a simple algebra with center $\mathfrak{k}$, and let $\mathfrak{K}$ be any splitting field of B which is normal separable over $\mathfrak{k}$. (We have just proved that every B has such a splitting field.) From Theorem 8.3A it follows at once that B is similar to an algebra A containing $\mathfrak{K}$ as a maximal subfield: $A^{\mathfrak{K}} = \mathfrak{K}$. Let $(\mathfrak{K} : \mathfrak{k}) = n$; by 7.3C, $(A : \mathfrak{k}) = n^2$. The Galois theory states that the set of all automorphisms of $\mathfrak{K}$ which leave elements of $\mathfrak{k}$ invariant form a group G of order n, and that $\mathfrak{k}$ is the set of all elements left invariant by all these automorphisms. If $\sigma \in G$, Corollary 7.2D shows that σ is induced by an inner automorphism of A, so that there is a $u_\sigma \in A$ such that $u_\sigma \alpha u_\sigma^{-1} = \alpha^\sigma$, that is,

$$(4.1) \qquad u_\sigma \alpha = \alpha^\sigma u_\sigma \quad \text{for all} \quad \alpha \in \mathfrak{K}.$$

(We denote the image of α under the automorphism σ by α^σ. Operators in the exponents are combined as if they were left operators; the map $(\alpha^\sigma)^\tau$ of α^σ under automorphism τ is written $\alpha^{\tau\sigma}$.)

Theorem 8.4A. If for each $\sigma \varepsilon G$ we choose a u_σ satisfying (4.1), then these n elements u_σ (considered as generators of a left $\mathfrak{k}$-space), are independent over $\mathfrak{K}$. Every element of A is uniquely expressible in the form

$$\sum_{\sigma \varepsilon G} \varkappa_\sigma u_\sigma, \quad \varkappa_\sigma \varepsilon \mathfrak{K}$$

Proof. Suppose the u_σ are not independent over $\mathfrak{K}$, and let

$$\alpha_\sigma u_\sigma + \alpha_\rho u_\rho + \cdots + \alpha_\tau u_\tau = 0 \tag{4.2}$$

be a shortest nontrivial relation, that is, a relation with fewest nonzero terms, in which not all the $\alpha_\sigma, \alpha_\rho, \cdots, \alpha_\tau$ are 0. Any such relation must contain at least two terms, for since u_σ has an inverse, $\alpha_\sigma u_\sigma = 0$ implies $\alpha_\sigma = 0$. Hence we can assume that α_σ and α_ρ are nonzero. If γ is any element of $\mathfrak{K}$, we get, by right multiplication by γ and left multiplication by $1/\tau^\sigma$:

$$\alpha_\sigma u_\sigma + \alpha_\rho (\gamma^\rho/\gamma^\sigma) u_\rho + \cdots + \alpha_\tau (\gamma^\tau/\gamma^\sigma) u_\tau = 0. \tag{4.3}$$

Now since ρ and σ are different automorphisms, we can choose $\gamma \varepsilon \mathfrak{K}$, so that $\gamma^\rho/\gamma^\sigma \neq 1$; then subtracting (4.3) from (4.2) gives

$$\alpha_\rho (1 - \gamma^\rho/\gamma^\sigma) u_\rho + \cdots + \alpha_\tau (1 - \gamma^\tau/\gamma^\sigma) u_\tau = 0, \tag{4.4}$$

which is a contradiction, since (4.4), having the term in u_σ missing, is a shorter nontrivial relation than (4.2).

Since $(A : \mathfrak{K}) = n$, the elements u_σ are a $\mathfrak{K}$-basis for A.

Choose a fixed set of u_σ, one for each automorphism. In order to describe A fully, we need only know the way in which the u_σ combine under multiplication; for if we know this we can compute all products of elements $\Sigma \alpha_\sigma u_\sigma$ by use of (4.1) and the distributive law.

If σ and τ are elements of G, (4.1) gives:

$$u_\sigma u_\tau \alpha = u_\sigma \alpha^\tau u_\tau = \alpha^{\sigma\tau} u_\sigma u_\tau,$$
$$u_{\sigma\tau} \alpha = \alpha^{\sigma\tau} u_{\sigma\tau}.$$

for all $\alpha \varepsilon \mathfrak{K}$. Thus $u_\sigma u_\tau \cdot u_{\sigma\tau}^{-1}$ is commutative with all $\alpha \varepsilon \mathfrak{K}$; since $A^{\mathfrak{K}} = \mathfrak{K}$, it is an element of $\mathfrak{K}$. We denote this element by $a_{\sigma,\tau}$, so that

$$u_\sigma u_\tau = a_{\sigma,\tau} u_{\sigma\tau}, \quad a_{\sigma,\tau} \varepsilon \mathfrak{K}. \tag{4.5}$$

Since the u_σ have inverses, $a_{\sigma,\tau} \neq 0$.

Clearly the structure of A is completely determined if $\mathfrak{K}$, $\mathfrak{k}$, and the constants $a_{\sigma,\tau}$ are known. (But one should note that a simple algebra A probably contains an infinite number of maximal separable splitting fields, and that for each field there is probably an infinite number of choices for the $a_{\sigma,\tau}$.)

We now change our point of view. Instead of starting with an algebra A, we choose a normal separable extension $\mathfrak{K}$, of degree n over $\mathfrak{k}$, introduce formally a set of elements u_σ, one for each $\sigma \in G$, form the left space generated by the u_σ over $\mathfrak{K}$, and define multiplication by

$$(4.1) \qquad u_\sigma \alpha = \alpha^\sigma u_\sigma,$$

$$(4.5) \qquad u_\sigma u_\tau = a_{\sigma,\tau} u_{\sigma\tau},$$

and by the distributive law, that is, we define

$$(4.6) \qquad (\Sigma\alpha_\sigma u_\sigma)(\Sigma\beta_\tau u_\tau) = \sum_{\sigma,\tau} (\alpha_\sigma u_\sigma)(\beta_\tau u_\tau) = \sum_{\sigma,\tau} (\alpha_\sigma \beta_\tau^\sigma a_{\sigma,\tau}) u_{\sigma\tau},$$

where the $a_{\sigma,\tau}$ are now an arbitrarily chosen set of elements of $\mathfrak{K}$. Is the resulting system a ring?

To prove the distributive law, it is enough to show that

$$\alpha_\rho u_\rho(\alpha_\sigma u_\sigma + \alpha_\tau u_\tau) = \alpha_\rho u_\rho \alpha_\sigma u_\sigma + \alpha_\rho u_\rho \alpha_\tau u_\tau.$$

and that we have the corresponding law with $\alpha_\rho u_\rho$ on the other side. These laws follow from (4.6) if $\sigma \neq \tau$; for $\sigma = \tau$ they follow from the fact that ρ is an additive automorphism:

$$\alpha_\rho u_\rho(\alpha u_\sigma + \beta u_\sigma) = \alpha_\rho u_\rho(\alpha + \beta)u_\sigma = \alpha_\rho(\alpha + \beta)^\rho u_\rho u_\sigma$$
$$= \alpha_\rho u_\rho \alpha u_\sigma + \alpha_\rho u_\rho \beta u_\sigma.$$

The associative law will hold if it holds for all cases of the form $(\alpha u_\rho \beta u_\sigma)\gamma u_\tau = \alpha u_\rho(\beta u_\sigma \gamma u_\tau)$. The two sides of this equation are equal to

$$\alpha\beta^\rho\gamma^{\rho\sigma}((u_\rho u_\sigma)u_\tau) \quad \text{and} \quad \alpha\beta^\rho\gamma^{\rho\sigma}(u_\rho(u_\sigma u_\tau)).$$

And since $(u_\rho u_\sigma)u_\tau = (a_{\rho,\sigma} u_{\rho\sigma})u_\tau = a_{\rho,\sigma} a_{\rho\sigma,\tau} u_{\rho\sigma\tau}$; $u_\rho(u_\sigma u_\tau) = u_\rho a_{\sigma,\tau} u_{\sigma\tau} = a^\rho_{\sigma,\tau} a_{\rho,\sigma\tau} u_{\rho\sigma\tau}$, it is evident that our system is associative if and only if the $a_{\sigma,\tau}$ satisfy

$$(4.7) \qquad a_{\rho,\sigma} a_{\rho\sigma,\tau} = a^\rho_{\sigma,\tau} a_{\rho,\sigma\tau} \quad \text{for all} \quad \rho,\sigma,\tau \in G.$$

A set of constants $a_{\sigma,\tau}$ which satisfy (4.7) is called a <u>factor set</u> of $\mathfrak{K}/\mathfrak{k}$ ($\mathfrak{K}/\mathfrak{k}$, read "$\mathfrak{K}$ over $\mathfrak{k}$," signifies $\mathfrak{K}$ is an extension field of $\mathfrak{k}$). A ring defined as above by means of a normal separable extension field $\mathfrak{K}/\mathfrak{k}$ and a factor set is called a <u>crossed product</u>, and is denoted by $(\mathfrak{K}/\mathfrak{k}, a_{\sigma,\tau})$. Our discussion proves

Theorem 8.4B. Every simple algebra is similar to a crossed product. Indeed, if A is simple with center $\mathfrak{k}$ and if $\mathfrak{K}/\mathfrak{k}$ is any normal separable splitting field for A, then for suitable $a_{\sigma,\tau}$, $A \cong (\mathfrak{K}/\mathfrak{k}, a_{\sigma,\tau})$.

Conversely, we have

Theorem 8.4C. Every crossed product $A = (\mathfrak{K}/\mathfrak{k}, a_{\sigma,\tau})$ is a simple algebra with center $\mathfrak{k}$; it has $\mathfrak{K}$ as splitting field.

Proof. We have already shown that A is an associative ring; it is an algebra of degree n^2 over $\mathfrak{k}$, since by construction the elements u_σ form a $\mathfrak{K}$-basis for A. To show that A is simple, we first prove that, if 1 stands for the identity automorphism of $\mathfrak{K}$, $e = a_{1,1}^{-1}u_1$ is unit element of A. For e is commutative with elements of $\mathfrak{K}$; furthermore, $eu_\sigma = a_{1,1}^{-1}a_{1,\sigma}u_\sigma$; $u_\sigma e = (a_{1,1}^{\sigma})^{-1}a_{\sigma,1}u_\sigma$, and the special cases

$$a_{1,1}a_{1,\sigma} = a_{1,\sigma}^{1}a_{1,\sigma}; \qquad a_{\sigma,1}a_{\sigma,1} = a_{1,1}^{\sigma}a_{\sigma,1}$$

of (4.7) show that $a_{1,1}^{-1}a_{1,\sigma} = 1$, and $(a_{1,1}^{\sigma})^{-1}a_{\sigma,1} = 1$; hence $eu_\sigma = u_\sigma e = u_\sigma$ and $e(\Sigma\alpha_\sigma u_\sigma) = \Sigma\alpha_\sigma u_\sigma$.

The subfield $e\mathfrak{K}$ of A is clearly isomorphic to $\mathfrak{K}$. It makes no difference in any of our formulas if we identify this subfield with $\mathfrak{K}$ (since $(e\alpha)u_\sigma = \alpha eu_\sigma = \alpha u_\sigma$ for any u_σ, any $\alpha \in \mathfrak{K}$) and we shall do so, considering $\mathfrak{K}$ as subring of A.

Let $\mathfrak{a}$ be any nonzero two-sided ideal of A and $a = \alpha_\rho u_\rho + \alpha_\sigma u_\sigma + \cdots + \alpha_\tau u_\tau$ be a shortest nonzero element of $\mathfrak{a}$. We again use the trick employed in proving 8.4A: if a contains more than one term, we can find γ in $\mathfrak{K}$ such that $\gamma^\sigma/\gamma^\rho \neq 1$; then $a - (1/\gamma^\rho)a\gamma = \alpha_\rho(1 - \gamma^\rho/\gamma^\rho)u_\rho + \alpha_\sigma(1 - \gamma^\sigma/\gamma^\rho)u_\sigma + \cdots + \alpha_\tau(1 - \gamma^\tau/\gamma^\rho)u_\tau \neq 0$ has at least one less term than a, and is in $\mathfrak{a}$. Then a must contain a nonzero element of one term, that is, an element αu_ρ, $\alpha \neq 0$. But such an element has an inverse:

$$(\alpha u_\rho)(u_{\rho^{-1}}a_{\rho,\rho^{-1}}^{-1}\alpha^{-1}a_{1,1}^{-1}) = \alpha a_{\rho,\rho^{-1}}u_1 a_{\rho,\rho^{-1}}^{-1}\alpha^{-1}a_{1,1}^{-1} = e.$$

Thus A can contain no two-sided ideal other than itself and (0).

If an algebra has a cyclic splitting field, its factor set can be written in a much simpler form. Suppose that $\mathfrak{K}/\mathfrak{k}$ is cyclic of degree n, and that σ is a fixed generator of its Galois group. If $\mathfrak{K}/\mathfrak{k}$ is a splitting field of the simple algebra A with center $\mathfrak{k}$, and $A^{\mathfrak{K}} = \mathfrak{K}$, choose $u_\sigma \in A$ such that

(4.8) $$u_\sigma \alpha = \alpha^\sigma u_\sigma \quad \text{for} \quad \alpha \in \mathfrak{K}.$$

Then, since $u_\sigma^\nu \alpha = \alpha^{(\sigma^\nu)} u_\sigma^\nu$, we can take

(4.9) $$\begin{cases} u_{\sigma\nu} = u_\sigma^\nu & \text{for} \quad \nu = 1, 2, \cdots, n-1 \\ u_1 = u_\sigma^0 = 1. \end{cases}$$

Since $u_\sigma^n \alpha = \alpha^{(\sigma^n)} u_\sigma^n = \alpha u_\sigma^n$, we see that u_σ^n is commutative with all elements of $\mathfrak{K}$; since it also is commutative with all u_σ, it is in the center of A:

(4.10) $$u_\sigma^n = a \in \mathfrak{k}.$$

Clearly, A is completely described by giving $\mathfrak{K}/\mathfrak{k}$, σ, and a. Thus we write A = $(\mathfrak{K}/\mathfrak{k}, \sigma, a)$. The product of two of the operators (4.9) is

(4.11) $$u_\sigma^\nu u_\sigma^\mu = \begin{cases} u_\sigma^{\nu+\mu} & \text{if} \quad \nu + \mu < n \\ a u_\sigma^{\nu+\mu-n} & \text{if} \quad \nu + \mu \geq n \end{cases}$$

(where ν, $\mu = 0 \cdots n-1$). So A = $(\mathfrak{K}/\mathfrak{k}, \sigma, a)$ is merely a special crossed product $(\mathfrak{K}/\mathfrak{k}, a_{\sigma,\tau})$, with

(4.12) $$a_{\sigma^\nu, \sigma^\mu} = \begin{cases} 1 & \text{if} \quad \nu + \mu < n \\ a & \text{if} \quad \nu + \mu \geq n. \end{cases}$$

If we begin with the cyclic field $\mathfrak{K}/\mathfrak{k}$, and define the system generated over $\mathfrak{K}$ by the n independent elements 1, u_σ, $u_\sigma^2 \cdots u_\sigma^{n-1}$, with the relations (4.8) and (4.10), choosing in (4.10) an arbitrary $a \neq 0$ in $\mathfrak{k}$, the resulting system is always a simple algebra. For, in view of our previous discussion, we have only to check that

$$u_\sigma^\lambda (u_\sigma^\mu u_\sigma^\nu) = (u_\sigma^\lambda u_\sigma^\mu) u_\sigma^\nu$$

when these products are computed by (4.12). This is clear, since, if we compute the product on either side, replace a by u_σ^n, and add exponents, we get in both cases $u_\sigma^{\lambda+\mu+\nu}$. Thus we have shown

<u>Theorem 8.4D</u>. <u>If a simple algebra A with center $\mathfrak{k}$ contains a maximal splitting field $\mathfrak{K}$ which is cyclic over $\mathfrak{k}$, then A = $(\mathfrak{K}/\mathfrak{k}, \sigma, a)$ for some a $\in \mathfrak{k}$. Conversely, for any a $\in \mathfrak{k}$, a $\neq$ 0, $(\mathfrak{K}/\mathfrak{k}, \sigma, a)$ is a simple algebra with center $\mathfrak{k}$ and splitting field $\mathfrak{K}$. $(\mathfrak{K}/\mathfrak{k}, \sigma, a) = (\mathfrak{K}/\mathfrak{k}, a_{\sigma,\tau})$ with $a_{\sigma,\tau}$ given by (4.12).</u>

Algebras containing a cyclic maximal splitting field are called <u>cyclic algebras</u>; their importance is shown by the fact that every simple algebra which is of finite degree over the rational field is a cyclic algebra.

Corollary 8.4E. *If ν is prime to the degree n of $\mathfrak{K}/\mathfrak{k}$, then $(\mathfrak{K}/\mathfrak{k}, \sigma, a) = (\mathfrak{K}/\mathfrak{k}, \sigma^{\nu}, a^{\nu})$.*

For σ^{ν} is also a generating automorphism, and $u_{\sigma}^{n} = a$ implies $(u_{\sigma}^{\nu})^{n} = a^{\nu}$.

We now investigate when two factor sets for a given field lead to the same simple algebra,

Theorem 8.4F. *$(\mathfrak{K}/\mathfrak{k}, a_{\sigma,\tau}) \cong (\mathfrak{K}/\mathfrak{k}, b_{\sigma,\tau})$ if and only if $\mathfrak{K}$ contains elements c_{σ} (one for each σ in the Galois group G) such that*

$$b_{\sigma,\tau} = \frac{c_{\sigma}c_{\tau}^{\sigma}}{c_{\sigma\tau}}\, a_{\sigma,\tau}. \tag{4.13}$$

If $a_{\sigma,\tau}$ is a factor set and $b_{\sigma,\tau}$ is given by (4.13), $b_{\sigma,\tau}$ is also a factor set. If $\mathfrak{K}/\mathfrak{k}$ is cyclic then $(\mathfrak{K}/\mathfrak{k}, \sigma, a) \cong (\mathfrak{K}/\mathfrak{k}, \sigma, b)$ if and only if $\mathfrak{K}$ contains an element c such that

$$b = aN_{\mathfrak{K}/\mathfrak{k}}c,$$

where $N_{\mathfrak{K}/\mathfrak{k}}c$ is defined to be $\Pi_{\sigma \varepsilon G}\; c^{\sigma}$.

(Factor sets related by an equation of form (4.13) are said to be *associated*.)

Proof. If $(\mathfrak{K}/\mathfrak{k}, a_{\sigma,\tau}) \cong (\mathfrak{K}/\mathfrak{k}, b_{\sigma,\tau})$, we can assume we have a single algebra A which contains subfields $\mathfrak{K}$ and $\mathfrak{K}'$ such that $A = (\mathfrak{K}/\mathfrak{k}, a_{\sigma,\tau}) = (\mathfrak{K}'/\mathfrak{k}, b'_{\sigma',\tau'})$, where σ' are the elements of the Galois group $\mathfrak{G}'$ of $\mathfrak{K}'/\mathfrak{k}$, and we have an isomorphic mapping of $\mathfrak{K}'$ onto $\mathfrak{K}$ and of $\mathfrak{G}'$ onto $\mathfrak{G}$ such that $\varkappa^{\sigma} \leftrightarrow \varkappa'^{\sigma'}$ and $b_{\sigma,\tau} \leftrightarrow b'_{\sigma',\tau'}$. $\mathfrak{K}'$ and $\mathfrak{K}$ may, of course, be different subfields.

But, by Corollary 7.2D, our isomorphism of $\mathfrak{K}'$ to $\mathfrak{K}$ can be extended to an inner automorphism of A; this inner automorphism will take $\mathfrak{K}'$ and the set of operators $v'_{\sigma'}$ which generate $(\mathfrak{K}'/\mathfrak{k}, b'_{\sigma',\tau'})$ into $\mathfrak{K}$ and a set of operators v_{σ} which generate a crossed product $(\mathfrak{K}/\mathfrak{k}, b_{\sigma,\tau})$. Accordingly, we can assume that $\mathfrak{K} = \mathfrak{K}'$: $A = (\mathfrak{K}/\mathfrak{k}, a_{\sigma,\tau}) = (\mathfrak{K}/\mathfrak{k}, b_{\sigma,\tau})$.

Thus A contains elements u_{σ}, v_{σ} such that

$$u_{\sigma}\alpha = \alpha^{\sigma}u_{\sigma} \qquad v_{\sigma}\alpha = \alpha^{\sigma}v_{\sigma} \tag{4.14}$$

$$u_{\sigma}u_{\tau} = a_{\sigma,\tau}u_{\sigma\tau} \qquad v_{\sigma}v_{\tau} = b_{\sigma,\tau}v_{\sigma\tau}. \tag{4.15}$$

From (4.14) we see that $v_{\sigma}u_{\sigma}^{-1}$ is in $A^{\mathfrak{K}} = \mathfrak{K}$; hence we can write $v_{\sigma} = c_{\sigma}u_{\sigma}$, where $c_{\sigma} \varepsilon \mathfrak{K}$. But then

$$(4.16)\quad v_\sigma v_\tau = c_\sigma u_\sigma c_\tau u_\tau = c_\sigma c_\tau^\sigma u_\sigma u_\tau = \frac{c_\sigma c_\tau^\sigma}{c_{\sigma\tau}} a_{\sigma,\tau} v_{\sigma\tau},$$

so that $b_{\sigma,\tau} = \frac{c_\sigma c_\tau^\sigma}{c_{\sigma\tau}} a_{\sigma,\tau}$.

Conversely, let $a_{\sigma,\tau}$ be any factor set, let $A = (\mathfrak{K}/\mathfrak{k}, a_{\sigma,\tau})$, and let u_σ be elements of A such that $u_\sigma\alpha = \alpha^\sigma u_\sigma$, $u_\sigma u_\tau = a_{\sigma,\tau} u_{\sigma\tau}$. Then, if c_σ are any set of nonzero elements of $\mathfrak{K}$, and we let $v_\sigma = c_\sigma u_\sigma$, the same computation (4.16) shows that $\mathfrak{K}/\mathfrak{k}$, together with the v_σ, generates the crossed product $(\mathfrak{K}/\mathfrak{k}, b_{\sigma,\tau})$, where $b_{\sigma,\tau}$ is given by (4.13).

Finally, let $\mathfrak{K}/\mathfrak{k}$ be cyclic of degree n and let the algebra $A = (\mathfrak{K}/\mathfrak{k}, \sigma, a)$ contain the element u_σ with $u_\sigma^n = a$ and $u_\sigma\alpha = \alpha^\sigma u_\sigma$. Then the above argument shows that $A \cong (\mathfrak{K}/\mathfrak{k}, \sigma, b)$ if and only if $\mathfrak{K}$ contains an element c such that $(cu_\sigma)^n = b$. But $(cu_\sigma)^n = cu_\sigma cu_\sigma \cdots cu_\sigma = cc^\sigma c^{\sigma^2} \cdots c^{\sigma^{n-1}} u_\sigma^n = aN_{\mathfrak{K}/\mathfrak{k}}c = b$.

<u>Corollary 8.4G.</u> $(\mathfrak{K}/\mathfrak{k}, a_{\sigma,\tau})$ <u>is a total matrix algebra if and only if there exist</u> $c_\sigma \in \mathfrak{K}$ <u>such that</u>

$$a_{\sigma,\tau} = \frac{c_\sigma c_\tau^\sigma}{c_{\sigma\tau}}.$$

<u>Proof.</u> In view of 8.4F, we need prove only that $(\mathfrak{K}/\mathfrak{k}, a_{\sigma,\tau}) \cong M_n$ when all $a_{\sigma,\tau} = 1$.

<u>Lemma 8.4H.</u> <u>If</u> $\mathfrak{K}$ <u>is any field and</u> $\sigma_1, \cdots \sigma_n$ <u>are different automorphisms of</u> $\mathfrak{K}$, <u>then these automorphisms, considered as linear functions of</u> $\mathfrak{K}$, <u>are linearly independent over</u> $\mathfrak{K}$, <u>that is, if</u> $\alpha_1, \cdots, \alpha_n$ <u>are in</u> $\mathfrak{K}$ <u>and</u>

$$\alpha_1\sigma_1(x) + \alpha_2\sigma_2(x) + \cdots + \alpha_n\sigma_n(x) = 0$$

<u>for all</u> $x \in \mathfrak{K}$, <u>then each</u> $\alpha_\nu = 0$.

(Here we write $\sigma(x)$ instead of our usual x^σ since this presents the ideas of the lemma more clearly.)

<u>Proof of 8.4H.</u> Suppose that

$$(4.17)\quad \alpha_1\sigma_1(x) + \alpha_2\sigma_2(x) + \cdots + \alpha_r\sigma_r(x) \equiv 0$$

is a shortest nontrivial relation. Since $\sigma_1(x)$ and $\sigma_2(x)$ are different, we can find $\gamma \in \mathfrak{K}$ such that $\sigma_1(\gamma) \neq \sigma_2(\gamma)$; then $\frac{1}{\sigma_1(\gamma)} (\alpha_1\sigma_1(\gamma x) + \alpha_2\sigma_2(\gamma x) + \cdots \alpha_r\sigma_r(\gamma x)) = 0$. But this gives

$$(4.18) \quad \alpha_1\sigma_1(x) + \alpha_2\frac{\sigma_2(\gamma)}{\sigma_1(\gamma)}\sigma_2(x) + \cdots + \alpha_r\frac{\sigma_r(\gamma)}{\sigma_1(\gamma)}\sigma_r(x) = 0,$$

and subtracting (4.18) from (4.17) gives a shorter non-trivial relation, which is a contradiction.

This lemma really belongs to the Galois theory.

Proof of 8.4G. Let $\mathfrak{K}/\mathfrak{k}$ be normal separable of degree n, consider $\mathfrak{K}$ as a vector space over $\mathfrak{k}$, and let M be the ring of all linear transformations of $\mathfrak{K}$ with respect to $\mathfrak{k}$; M is isomorphic to the $n \times n$ total matrix algebra over $\mathfrak{k}$.

If $\alpha \varepsilon \mathfrak{K}$, the transformation $\overline{\alpha}(x) = \alpha \cdot x$ is linear with respect to $\mathfrak{k}$; these transformations $\overline{\alpha}(x)$ form a subring $\overline{\mathfrak{K}}$ of M which is isomorphic to $\mathfrak{K}$. For each of the n automorphisms σ of $\mathfrak{K}$ which leaves elements of $\mathfrak{k}$ invariant, the transformation $u_\sigma(x) = \sigma(x)$ is in M, since $\sigma(\alpha + \beta) = \sigma(\alpha) + \sigma(\beta)$ and, if $a \varepsilon \mathfrak{k}$, $\sigma(a\alpha) = a\sigma(\alpha)$. By Lemma 8.4H the transformations $u_\sigma(x)$ are linearly independent over $\mathfrak{K}$; hence, since $(M : \overline{\mathfrak{K}}) = n$, $\overline{\mathfrak{K}}$ and the u_σ generate the whole ring M. But, since $\sigma(\alpha x) = \alpha^\sigma \cdot \sigma(x)$, we have

$$u_\sigma\overline{\alpha} = (\overline{\alpha^\sigma})u_\sigma.$$

Since $u_\sigma u_\tau = u_{\sigma\tau}$, $\overline{\mathfrak{K}}$ and the u_σ form a factor set, with $a_{\sigma,\tau} = 1$.

5. FORMALISM OF CROSSED PRODUCTS

If A and B are two crossed products with respect to $\mathfrak{K}/\mathfrak{k}$, we shall show that $A \times_{\mathfrak{k}} B$ is also such a crossed product. Furthermore, if $\mathfrak{k}'$ is an extension field of $\mathfrak{k}$, $A \times_{\mathfrak{k}} \mathfrak{k}'$ is a crossed product with respect to $\mathfrak{K}\mathfrak{k}'/\mathfrak{k}'$. Finally, if $\mathfrak{L}/\mathfrak{k}$ is normal separable, and includes $\mathfrak{K}/\mathfrak{k}$, then A is similar to some crossed product with respect to $\mathfrak{L}/\mathfrak{k}$. In all three cases the factor sets of the new crossed products can be computed from the original factor sets by simple rules. These rules are useful in studying the relation between theory of algebras and algebraic number theory.

Theorem 8.5A. Let $A = (\mathfrak{K}/\mathfrak{k}, a_{\sigma,\tau})$ and $B = (\mathfrak{K}/\mathfrak{k}, b_{\sigma,\tau})$. Then, if $c_{\sigma,\tau} = a_{\sigma,\tau} b_{\sigma,\tau}$, the numbers $c_{\sigma,\tau}$ also form a factor set.

$$A \times_{\mathfrak{k}} B \sim (\mathfrak{K}/\mathfrak{k}, c_{\sigma,\tau}).$$

For $\mathfrak{K}/\mathfrak{k}$ cyclic $(\mathfrak{K}/\mathfrak{k}, \sigma, a) \times_{\mathfrak{k}} (\mathfrak{K}/\mathfrak{k}, \sigma, b) = (\mathfrak{K}/\mathfrak{k}, \sigma, ab)$.

Proof. By our assumptions A contains a subfield $\mathfrak{K}' \cong \mathfrak{K}$, and B a field $\mathfrak{K}'' \cong \mathfrak{K}$. We shall suppose that we have a fixed isomorphism $\alpha' \leftrightarrow \alpha''$ of $\mathfrak{K}'$ to $\mathfrak{K}''$, and that we have defined the elements of the group G as automorphisms of both fields $\mathfrak{K}'$ and $\mathfrak{K}''$ in such a manner that, if $\alpha' \leftrightarrow \alpha''$, then $\alpha'^{\sigma} \leftrightarrow \alpha''^{\sigma}$ for all $\sigma \in G$.

Let $\mathfrak{K}' = \mathfrak{k}(\alpha')$, $\mathfrak{K}'' = \mathfrak{k}(\alpha'')$, where $\alpha' \leftrightarrow \alpha''$, let α satisfy the irreducible equation $f(x) = 0$, with coefficients in $\mathfrak{k}$, and leading coefficient 1, and consider the following element of the commutative subring $\mathfrak{K}' \times_{\mathfrak{k}} \mathfrak{K}''$ of $A \times_{\mathfrak{k}} B$:

$$e = \varphi(\alpha', \alpha'') = \frac{\prod_{\sigma \neq 1} (\alpha' - \alpha''^{\sigma})}{f'(\alpha')} . \tag{5.1}$$

Since the numerator is a polynomial of degree n-1 in α, with nonzero coefficients in $\mathfrak{K}''$, it follows from the definition of $\times_{\mathfrak{k}}$ that $e \neq 0$.

Since $f(x) = \prod_{\sigma \in G} (x - \alpha'^{\sigma}) = \prod_{\sigma \in G} (x - \alpha''^{\sigma})$, we have $e(\alpha' - \alpha'') = \frac{f(\alpha')}{f'(\alpha')} = 0$. Hence $e\alpha' = e\alpha''$. But this means that $e\alpha'^2 = e\alpha''\alpha' = e\alpha''^2, \cdots, e\alpha'^{\nu} = e\alpha''$. Now, since every element β' of $\mathfrak{K}'$ is expressible as a polynomial in α' with coefficients in $\mathfrak{k}$, we have

$$e\beta' = e\beta'' \tag{5.2}$$

for any $\beta' \in \mathfrak{K}'$, $\beta' \leftrightarrow \beta'' \in \mathfrak{K}''$.

Let $\tau \in G$ and consider the element $e' = \varphi(\alpha'^{\tau}, \alpha''^{\tau})$. We can write e' as a polynomial in α'^{τ} with coefficients in $\mathfrak{K}''$; since e is commutative with all elements of $\mathfrak{K}' \times_{\mathfrak{k}} \mathfrak{K}''$, (5.2) shows that

$$ee' = e\varphi(\alpha'^{\tau}, \alpha''^{\tau}) = e \frac{\prod_{\sigma \neq 1} (\alpha'^{\tau} - \alpha''^{\sigma\tau})}{f'(\alpha'^{\tau})} =$$

$$e \frac{\prod_{\sigma \neq 1} (\alpha'^{\tau} - \alpha'^{\sigma\tau})}{f'(\alpha'^{\tau})} = e \frac{f'(\alpha'^{\tau})}{f'(\alpha'^{\tau})} = e.$$

Thus $ee' = e \cdot \varphi(\alpha'^{\tau}, \alpha''^{\tau}) = e$ for all τ. Since the same reasoning could be applied to show that $ee' = e'$, we have

$$\begin{aligned} &e^2 = e \\ &e = \varphi(\alpha', \alpha'') = \varphi(\alpha'^{\tau}, \alpha''^{\tau}) \quad \text{for all} \quad \tau \in G. \end{aligned} \tag{5.3}$$

Let $A \times_{\mathfrak{k}} B = C$, and consider the algebra eCe. This subalgebra is generated over $\mathfrak{k}$ by $e\mathfrak{K}'e$, $e\mathfrak{K}''e$, and the elements $eu'_{\sigma}v''_{\pi}e$, where u' and v'' are elements of A and B, respectively, which satisfy

$$u'_\sigma \alpha' = \alpha'^\sigma u'_\sigma \qquad v''_\sigma \alpha'' = \alpha''^\sigma v''_\sigma$$
$$u'_\sigma u'_\tau = a'_{\sigma,\tau} u'_{\sigma\tau} \qquad v''_\sigma v''_\tau = b''_{\sigma,\tau} v''_{\sigma\tau}.$$

But $e\beta'' = e\beta'$ for all $\beta'' \in \mathfrak{K}''$; hence we can dispense with $e\mathfrak{K}''e$. Furthermore,

$$\begin{aligned} eu'_\sigma v''_\tau e &= eu'_\sigma v''_\tau \varphi(\alpha', \alpha'') = eu'_\sigma \varphi(\alpha', \alpha''^\tau) v''_\tau \\ &= e\varphi(\alpha'^\sigma, \alpha''^\tau) u'_\sigma v''_\tau \\ &= \begin{cases} eu'_\sigma v''_\sigma & \text{if} \quad \sigma = \tau \\ 0 & \text{if} \quad \sigma \neq \tau, \end{cases} \end{aligned}$$

(For it is evident from the definition (5.1) that $e\varphi(\alpha'^\sigma, \alpha''^\tau) = 0$ if $\sigma \neq \tau$—for then a factor of the numerator will be 0.)

Thus eCe is generated over $\mathfrak{k}$ by the field $\bar{\mathfrak{K}} = e\mathfrak{K}'e$ and the elements $\bar{u}_\sigma = eu'_\sigma v''_\sigma e$. If $\bar{\beta} \in \bar{\mathfrak{K}}$,

$$\begin{aligned} \bar{u}_\sigma \bar{\beta} &= eu'_\sigma v''_\sigma e^2 \beta' e = eu'_\sigma \beta' v''_\sigma e \\ &= e\beta'^\sigma u'_\sigma v''_\sigma e \\ &= \bar{\beta}^\sigma \bar{u}_\sigma. \end{aligned}$$

And

$$\begin{aligned} \bar{u}_\sigma \bar{u}_\tau &= eu'_\sigma v''_\sigma eu'_\tau v''_\tau e = eu'_\sigma v''_\sigma u'_\tau v''_\tau e^2 \\ &= eu'_\sigma u'_\tau v''_\sigma v''_\tau e = ea'_{\sigma,\tau} u'_{\sigma\tau} b''_{\sigma,\tau} v''_{\sigma\tau} e \\ &= ea'_{\sigma,\tau} eb''_{\sigma,\tau} e\bar{u}_{\sigma\tau} \\ &= \bar{c}_{\sigma,\tau} \bar{u}_{\sigma\tau}. \end{aligned}$$

Hence $eCe \cong (\mathfrak{K}/\mathfrak{k}, c_{\sigma,\tau})$.

To prove 8.5A it suffices to prove

<u>Lemma 8.5B</u>. <u>If</u> C <u>is any simple ring and</u> e <u>is an idempotent element contained in</u> C, <u>then</u> eCe ~ C.

<u>Proof of Lemma</u>. eCe is a simple ring. For if $\mathfrak{a}$ is a two-sided ideal in eCe, then $C\mathfrak{a}C$ is a two-sided ideal in C, hence $C\mathfrak{a}C$ is either (0) or C. But since $\mathfrak{a} = eCe\,\mathfrak{a}\,eCe = e(C\mathfrak{a}C)e$, this implies that $\mathfrak{a}$ is eCe or (0); therefore eCe has no proper two-sided ideal. To show that eCe has minimum condition, let $\mathfrak{l}_1 \supseteq \mathfrak{l}_2, \cdots, \supseteq \mathfrak{l}_\nu \cdots$ be a descending chain of ideals in eCe; then $C\mathfrak{l}_1 \supseteq C\mathfrak{l}_2 \supseteq \cdots$ is a descending chain of ideals in C, so that from some point on the $C\mathfrak{l}_i$ are equal. But $C\mathfrak{l}_i = C\mathfrak{l}_{i+1}$ implies $\mathfrak{l}_i = eC\mathfrak{l}_i = \mathfrak{l}_{i+1}$.

Now let $eCe \cdot e'$ be a minimal left ideal in eCe. By Theorem 5.4A, $e'(eCe)e'$ is a sfield; since e is unit for eCe, $e'(eCe)e' = e'Ce'$. By the Wedderburn Theorem 5.1A and §5.6 it follows that $eCe \sim e'Ce'$.

On the other hand, Theorem 5.4A shows that, since $e'Ce'$ is a sfield, Ce' is a minimal C-ideal; hence, by the same argument,

$$C \sim e'Ce' \sim eCe.$$

This proves Lemma 8.5B and Theorem 8.5A. (The statement about cyclic algebras in 8.5A follows at once, by writing $(\mathfrak{K}/\mathfrak{k}, \sigma, a) = (\mathfrak{K}/\mathfrak{k}, \sigma, a_{\sigma,\tau})$ with $a_{\sigma,\tau}$, as in Theorem 8.4D.)

Clearly the factor sets for a given normal separable $\mathfrak{K}/\mathfrak{k}$ form a group, if we define the product of $a_{\sigma,\tau}$ and $b_{\sigma,\tau}$ to be the factor set $c_{\sigma,\tau}$ of Theorem 8.5A. Theorems 8.5A and 8.4G give at once

Corollary 8.5C. *If $\mathfrak{K}/\mathfrak{k}$ is normal separable, then the group (under $\times_{\mathfrak{k}}$) of all classes of equivalent simple algebras split by $\mathfrak{K}$ and having center $\mathfrak{k}$ is isomorphic to the group of all factors set for $\mathfrak{K}/\mathfrak{k}$ modulo the subgroup of all factor sets of form* $a_{\sigma,\tau} = \dfrac{c_\sigma c_\tau^\sigma}{c_{\sigma\tau}}$.

Theorem 8.5D. *Let $\mathfrak{K}/\mathfrak{k}$ be normal separable with Galois group G, let $\mathfrak{k}'$ be any field containing $\mathfrak{k}$, and let G' be the group of all elements of G which leave all elements of $\mathfrak{K} \cap \mathfrak{k}'$ invariant. Then G' can be identified with the Galois group of $\mathfrak{K}\mathfrak{k}'/\mathfrak{k}'$. If $A = (\mathfrak{K}/\mathfrak{k}, a_{\sigma,\tau})$, then the set of elements $a_{\sigma',\tau'}$, obtained by letting σ' and τ' run through G' only, forms a factor set for $\mathfrak{K}\mathfrak{k}'/\mathfrak{k}'$, and*

$$A \times_{\mathfrak{k}} \mathfrak{k}' \sim (\mathfrak{K}\mathfrak{k}'/\mathfrak{k}', a_{\sigma',\tau'}).$$

If $\mathfrak{K}/\mathfrak{k}$ is cyclic with Galois group G generated by σ, and σ^ν generates the subgroup of G leaving $\mathfrak{K} \cap \mathfrak{k}'$ invariant, then

$$(\mathfrak{K}/\mathfrak{k}, \sigma, a) \times_{\mathfrak{k}} \mathfrak{k}' \sim (\mathfrak{K}\mathfrak{k}'/\mathfrak{k}', \sigma^\nu, a).$$

Proof. Whenever we write a product, of the form $\mathfrak{K}\mathfrak{k}'$, of two fields, it is assumed that $\mathfrak{K}$ and $\mathfrak{k}'$ are subfields of some larger field Ω; $\mathfrak{K}\mathfrak{k}'$ then stands for the smallest subfield of Ω containing $\mathfrak{K}$ and $\mathfrak{k}'$. (Note that there are no restrictions on the field $\mathfrak{k}'$; it can be of infinite degree over $\mathfrak{k}$.)

It is evident from Galois theory that the Galois group of $\mathfrak{K}\mathfrak{k}'/\mathfrak{k}'$ is formed by extending the elements of G' to automorphisms of $\mathfrak{K}\mathfrak{k}'$.*

Case I. $\mathfrak{k}' \subseteq \mathfrak{K}$. In this case we have, by 7.3C (1),

$$A \times_{\mathfrak{k}} \mathfrak{k}' \sim A^{\mathfrak{k}'}.$$

To compute $A^{\mathfrak{k}'}$ we note that every element of A can be written in the form

$$\xi = \sum_{\sigma \varepsilon G} \alpha_\sigma u_\sigma,$$

where the $\alpha_\sigma \ \varepsilon \ \mathfrak{K}$ and the u_σ are a set of elements such that $u_\sigma \alpha = \alpha^\sigma u_\sigma$ and $u_\sigma u_\tau = a_{\sigma,\tau} u_{\sigma\tau}$. The element ξ will be in $A^{\mathfrak{k}'}$ if and only if $\xi\beta - \beta\xi = 0$ for all $\beta \ \varepsilon \ \mathfrak{k}'$; that is, if and only if

$$\sum_{\sigma \varepsilon G} (\beta - \beta^\sigma) \alpha_\sigma u_\sigma = 0$$

for all $\beta \ \varepsilon \ \mathfrak{k}'$. But since the u_σ are linearly independent over $\mathfrak{K}$, this can be true if and only if $\alpha_\sigma = 0$, except when $\sigma \ \varepsilon \ G'$. Thus $A^{\mathfrak{k}'}$ consists exactly of the elements $\Sigma_{\sigma' \varepsilon G'} \alpha_{\sigma'} u_{\sigma'}$. Since the automorphisms σ' range exactly over the Galois group G' of $\mathfrak{K}/\mathfrak{k}'$, these elements generate exactly the crossed product $(\mathfrak{K}/\mathfrak{k}', a_{\sigma',\tau'})$. Since the $u_{\sigma'}$ are elements of the associative ring A, their products satisfy the associative law; hence the $a_{\sigma',\tau'}$ satisfy (4.7).

Case II. $\mathfrak{K} \cap \mathfrak{k}' = \mathfrak{k}$. Here $\mathfrak{K} \times_{\mathfrak{k}} \mathfrak{k}'$ and $\mathfrak{K}\mathfrak{k}'$ are isomorphic; for $(\mathfrak{K} : \mathfrak{k}) = (\mathfrak{K}\mathfrak{k}' : \mathfrak{k}')$, so that both rings can be represented as the set of all linear combinations, with elements in $\mathfrak{k}'$, of a set of linearly independent generators of $\mathfrak{K}$ over $\mathfrak{k}$, with multiplication rules for the basis elements the same as for multiplication in $\mathfrak{K}$. (If $\mathfrak{K} \cap \mathfrak{k}'$ is larger than $\mathfrak{k}$, this is no longer true, since $(\mathfrak{K}\mathfrak{k}' : \mathfrak{k}') < (\mathfrak{K} \times_{\mathfrak{k}} \mathfrak{k}' : \mathfrak{k}')$. In this case it is easy to show that $\mathfrak{K} \times_{\mathfrak{k}} \mathfrak{k}'$ is a direct sum of fields, each isomorphic to $\mathfrak{K}\mathfrak{k}'$. But that fact is not needed here.)

Accordingly, in this case, $A \times_{\mathfrak{k}} \mathfrak{k}'$ contains the normal subfield $\mathfrak{K}\mathfrak{k}'/\mathfrak{k}'$, whose Galois group is still G. Hence we can write $A \times_{\mathfrak{k}} \mathfrak{k}'$ as a crossed product

*Every automorphism of $\mathfrak{K}\mathfrak{k}'/\mathfrak{k}'$ induces an automorphism of $\mathfrak{K}$ which leaves $\mathfrak{K} \cap \mathfrak{k}'$ invariant; thus every automorphism of $\mathfrak{K}\mathfrak{k}'/\mathfrak{k}'$ is an extension of an automorphism in G'. Conversely, let σ leave $\mathfrak{K} \cap \mathfrak{k}'$ invariant and let $\mathfrak{K}\mathfrak{k}' = \mathfrak{k}'(\alpha)$, where $\alpha \ \varepsilon \ \mathfrak{K}$ satisfies an equation which is irreducible in $\mathfrak{k}'$ and has coefficients in $\mathfrak{K} \cap \mathfrak{k}'$. Since α^σ satisfies the same equation, σ can be extended to an automorphism of $\mathfrak{K}\mathfrak{k}'/\mathfrak{k}'$.

$$A \times_{\mathfrak{k}} \mathfrak{k}' = (\mathfrak{K}\mathfrak{k}'/\mathfrak{k}', a_{\sigma,\tau}),$$

using the same generators u_σ that were used in the original expression of A as crossed product.

To get the general case, we note that, if $\mathfrak{k}'' = \mathfrak{k}' \cap \mathfrak{K}$, we have, by 8.3F, $A \times_{\mathfrak{k}} \mathfrak{k}' \cong (A \times_{\mathfrak{k}} \mathfrak{k}'') \times_{\mathfrak{k}''} \mathfrak{k}'$; we then apply case I to $\times_{\mathfrak{k}}$ and case II to $\times_{\mathfrak{k}''}$.

Finally, if $\mathfrak{K}/\mathfrak{k}$ is cyclic of degree n, and σ^ν is the smallest power of a generating automorphism σ leaving elements of $\mathfrak{K} \cap \mathfrak{k}'$ invariant, then $A = (\mathfrak{K}/\mathfrak{k}, \sigma, a)$ is a simple algebra with center $\mathfrak{k}$, generated over $\mathfrak{K}$ by powers of u_σ, where $u_\sigma^n = a$. By the argument used in proving our theorem for general crossed products, we see that $A \times_{\mathfrak{k}} \mathfrak{k}'$ is similar to an algebra with center $\mathfrak{k}'$, generated by $\mathfrak{K}\mathfrak{k}'$ and powers of u_σ^ν. But $\mathfrak{K}\mathfrak{k}'/\mathfrak{k}'$ is of degree n/ν and $(u_\sigma^\nu)^{n/\nu} = a$. Hence we have

$$A \times_{\mathfrak{k}} \mathfrak{k}' \sim (\mathfrak{K}\mathfrak{k}'/\mathfrak{k}', \sigma^\nu, a).$$

Theorem 8.5E. *Let $\Omega \supset \mathfrak{K} \supset \mathfrak{k}$, where both Ω and $\mathfrak{K}$ are normal separable over $\mathfrak{k}$. Let G be the Galois group of $\Omega/\mathfrak{k}$ and H the Galois group of $\Omega/\mathfrak{K}$, and identify the Galois group of $\mathfrak{K}/\mathfrak{k}$ with the factor group G/H. Let $A = (\mathfrak{K}/\mathfrak{k}, a_{\bar\sigma,\bar\tau})$ be a simple algebra ($\bar\sigma$, $\bar\tau$ run through G/H).*

Then A can also be represented as a crossed product involving the extension field Ω; in fact, $A \sim (\Omega/\mathfrak{k}, b_{\sigma,\tau})$ with

$$b_{\sigma,\tau} = a_{\bar\sigma,\bar\tau},$$

where $\bar\sigma$, $\bar\tau$ are the residue classes of σ and τ, respectively, in G/H.

If $\mathfrak{K}/\mathfrak{k}$ and $\Omega/\mathfrak{k}$ are cyclic, and if σ generates G and σ^μ generates H, our rule reduces to

$$(\mathfrak{K}/\mathfrak{k}, \sigma, a) \sim (\Omega/\mathfrak{k}, \sigma, a^\mu).$$

Proof. Set $(\Omega : \mathfrak{k}) = m$ and form $A' = A \times_{\mathfrak{k}} M$, where M is an $m \times m$ total matrix algebra over $\mathfrak{k}$. Clearly A' contains Ω as maximal splitting field (for $(A' : \mathfrak{k}) = m^2(A : \mathfrak{k}) = m^2(\mathfrak{K} : \mathfrak{k})^2 = (\Omega : \mathfrak{k})^2$). We shall find a representation of A' as a crossed product with Ω.

First we construct, by the method of Theorem 3.2A, a subring of A' which is isomorphic to Ω; let $\omega_1 \cdots \omega_m$ be an independent basis for Ω over $\mathfrak{K}$, and form the vector

$$\omega = (\omega_1, \omega_2, \cdots \omega_m).$$

For each $\alpha \in \Omega$ there is a uniquely determined matrix $M(\alpha)$, with elements in $\mathfrak{K}$, satisfying

(5.4) $$\alpha \cdot \omega = \omega M(\alpha).$$

These matrices $M(\alpha)$ form a subring Ω' of A'; clearly $\Omega' \cong \Omega$.

If τ is any element of G, and $\omega^\tau = (\omega_1^\tau, \omega_2^\tau, \dots \omega_m^\tau)$, there is similarly a uniquely defined matrix P_τ, with elements in $\mathfrak{K}$, satisfying

(5.5) $$\omega^\tau = \omega P_\tau.$$

If $\sigma \varepsilon G$, and we apply σ to (5.5), we get $\omega^{\sigma\tau} = \omega^\sigma P_\tau^\sigma = \omega P_\sigma P_\sigma^\tau$. Since, by definition (5.5), $\omega^{\sigma\tau} = \omega P_{\sigma\tau}$, we have

(5.6) $$P_{\sigma\tau} = P_\sigma P_\tau^\sigma.$$

If we apply σ to (5.4), we get $\alpha^\sigma\omega^\sigma = \omega^\sigma M^\sigma(\alpha) = \omega P_\sigma M^\sigma(\alpha)$ (where, of course, M^σ is the matrix obtained by replacing each element of M by its map under the automorphism σ). On the other hand, $\alpha^\sigma\omega^\sigma = \alpha^\sigma\omega P_\sigma = \omega M(\alpha^\sigma)P_\sigma$. We thus find

(5.7) $$P_\sigma M^\sigma(\alpha) = M(\alpha^\sigma)P_\sigma.$$

We observe that the correspondences $M(\alpha) \to M^\sigma(\alpha)$ form a group of automorphisms of Ω' which leaves invariant exactly the matrices $M(\varkappa) = \varkappa I$, where $\varkappa$ is in $\mathfrak{k}$. Thus the set of these automorphisms of Ω' forms the Galois group of Ω' over $\mathfrak{k}$. But the correspondences $M(\alpha) \to M^\sigma(\alpha)$ do <u>not</u> induce all automorphisms of Ω' over $\mathfrak{k}$; in fact, since the elements of M are in $\mathfrak{K}$, $M^\sigma(\alpha)$ really is completely determined by the residue class $\bar\sigma$ of σ in G/H. We thus write $M^{\bar\sigma}(\alpha)$ instead of $M^\sigma(\alpha)$. The same remarks apply to the matrices P_τ; we write $P_\tau^{\bar\sigma} = P_\tau^\sigma$, since P_τ^σ is completely known when $\bar\sigma$ is known.

Now to represent A' as a crossed product, let $u_{\bar\sigma}$ be elements of A satisfying

$$u_{\bar\sigma}\alpha = \alpha^{\bar\sigma}u_{\bar\sigma} \quad \text{for } \alpha \varepsilon \mathfrak{K}$$

$$u_{\bar\sigma}u_{\bar\tau} = a_{\bar\sigma,\bar\tau}u_{\bar\sigma\bar\tau},$$

and define u'_σ to be the element $P_\sigma u_{\bar\sigma}$ of A'. Then (5.7) gives

$$u'_\sigma M(\alpha) = P_\sigma u_{\bar\sigma}M(\alpha) = P_\sigma M^{\bar\sigma}(\alpha)u_\sigma = M(\alpha^\sigma)P_\sigma u_{\bar\sigma},$$

that is,

(5.8) $$u'_\sigma M(\alpha) = M(\alpha^\sigma)u'_\sigma.$$

And (5.6) gives

$$u'_\sigma u'_\tau = P_\sigma u_{\bar\sigma}P_\tau u_{\bar\tau} = P_\sigma P_\tau^{\bar\sigma}u_{\bar\sigma}u_{\bar\tau} = a_{\bar\sigma,\bar\tau}P_\sigma P_\tau^\sigma u_{\bar\sigma\bar\tau}.$$

So we have

$$u'_{\sigma}u'_{\tau} = a_{\bar{\sigma},\bar{\tau}}u'_{\sigma\tau}, \tag{5.9}$$

when $\bar{\sigma}$, $\bar{\tau}$ are the residue classes of σ, τ in G/H.

But (5.8) and (5.9) show that A' is generated by Ω' and the u'_{σ}, and is thus equal to the crossed product $(\Omega'/\mathfrak{k}, a_{\sigma,\tau})$.

6. EXPONENTS OF SIMPLE ALGEBRAS

The exponent of A is defined as the smallest positive integer x such that $A^x \sim 1$, where, if $\mathfrak{k}$ is the center of A, A^x means the $\times_{\mathfrak{k}}$ product of x algebras, all isomorphic to A. That is, the exponent of A is the period of the class of algebras similar to A in the group of all classes of simple algebras with center $\mathfrak{k}$. We investigate the number x; our first theorem shows that every A does have a finite exponent.

Theorem 8.6A. If A is simple with center $\mathfrak{k}$ and $(A : \mathfrak{k}) = n^2$, then $A^n \sim \mathfrak{k}$: the exponent of A divides n.

Proof. Since the dimension of any simple algebra over its center is a multiple of the dimension of its division algebra component, it suffices to prove the theorem for division algebras. Let D be a division algebra with center $\mathfrak{k}$, and let $(D : \mathfrak{k}) = n^2$. Let $\mathfrak{K}$ be a splitting field of D which is normal separable over $\mathfrak{k}$. By Theorem 8.3A we see that we can find an integer r such that $\mathfrak{K}$ is a maximal subfield of the algebra

$$B = D \times_{\mathfrak{k}} M_r.$$

Then $(\mathfrak{K} : \mathfrak{k}) = \sqrt{(B : \mathfrak{k})} = nr$. Furthermore, for a suitable factor set $a_{\sigma,\tau}$,

$$B = (\mathfrak{K}/\mathfrak{k}, a_{\sigma,\tau}). \tag{6.1}$$

Let the elements u_{σ} be the usual set of generators of B over $\mathfrak{K}$ used in the expression of B as such a crossed product.

Considered as a space with itself as left operator domain, B is a direct sum of r minimal left ideals. Let $\mathfrak{l}$ be one of these ideals; since our operator domain B contains $\mathfrak{K}$, we can consider $\mathfrak{l}$ as a right $\mathfrak{K}$-space. Since $(\mathfrak{l} : D) = r$, we see that $(\mathfrak{l} : \mathfrak{k}) = n^2 r$, so that

$$(\mathfrak{l} : \mathfrak{K}) = n. \tag{6.2}$$

Let $\lambda_1, \cdots, \lambda_n$ be an independent basis for $\mathfrak{X}$ over $\mathfrak{K}$, and form the column vector

$$\Lambda = \begin{bmatrix} \lambda_1 \\ \lambda_2 \\ \cdot \\ \cdot \\ \cdot \\ \lambda_n \end{bmatrix}$$

Then for each u_τ there is a uniquely defined matrix U_τ, with elements in $\mathfrak{K}$, satisfying the equation

$$(6.3) \qquad u_\tau \Lambda = U_\tau \Lambda .$$

These matrices U_τ are not part of a matrix representation of B; but multiplying (6.3) on the left by u_σ, we get

$$\begin{aligned} u_\sigma u_\tau \Lambda &= u_\sigma U_\tau \Lambda = U_\tau^\sigma u_\sigma \Lambda = U_\tau^\sigma U_\sigma \Lambda \\ &= a_{\sigma,\tau} u_{\sigma\tau} \Lambda \qquad\quad = a_{\sigma,\tau} U_{\sigma\tau} \Lambda . \end{aligned}$$

Hence we have

$$(6.4) \qquad a_{\sigma,\tau} U_{\sigma\tau} = U_\tau^\sigma U_\sigma .$$

Taking determinants of both sides of (6.4), remembering that the U_σ are n x n matrices, and setting $c_\sigma = |U_\sigma|$, we get

$$(6.5) \qquad a_{\sigma,\tau}^n = \frac{c_\tau^\sigma c_\sigma}{c_{\sigma\tau}} .$$

But this means, by Theorem 8.4F, that $(\mathfrak{K}/\mathfrak{k}, a_{\sigma,\tau}^n) \sim 1$. Since, by Theorem 8.5A, $A^n \sim (\mathfrak{K}/\mathfrak{k}, a_{\sigma,\tau}^n)$, we have proved our Theorem 8.6A.

If the center $\mathfrak{k}$ of D is an algebraic number field, the exponent of D is exactly $\sqrt{(D : \mathfrak{k})}$. But the proof of this fact requires considerable algebraic number theory. For general algebras one can prove only

Theorem 8.6B. *If a prime* p *divides* $(D : \mathfrak{k})$, *where* D *is a division algebra and* $\mathfrak{k}$ *its center, then* p *divides the exponent of* D.

Proof. This theorem rests directly on the theorem from the theory of finite groups, which states that, if G is any finite group, and if p^ν is the highest power of p dividing the order of G, then G contains a subgroup (called a p-Sylow group) whose order is exactly p^ν. A nice proof of the theorem can be found in L. C. Mathewson, *Elementary Theory of Finite Groups*, (New York, 1930), p. 82, or in H. Zassenhaus, *Lehrbuch der Gruppentheorie* (Leipzig, 1937), p. 99.

Suppose that our theorem is false, that is, that we have a division algebra D with center $\mathfrak{k}$ such that a prime p divides $(D : \mathfrak{k})$ but is prime to the exponent of D. Let $\mathfrak{K}$ be a splitting field of D which is normal separable over $\mathfrak{k}$, and let G be the Galois group of $\mathfrak{K}/\mathfrak{k}$. Let S be a p-Sylow subgroup of G, and let $\mathfrak{k}'$ be the sufield of $\mathfrak{K}$ whose elements are left invariant by S. Then the degree $(\mathfrak{K} : \mathfrak{k}')$ is equal to the order of S and is thus a power of p, and $(\mathfrak{k}' : \mathfrak{k}) = (\mathfrak{K} : \mathfrak{k})/(\mathfrak{K} : \mathfrak{k}')$, and is thus prime to p.

Form the algebra $A = D \times_{\mathfrak{k}} \mathfrak{k}'$. A is a simple algebra with center $\mathfrak{k}'$, and it has $\mathfrak{K}$ as splitting field; hence, by Corollary 8.3C, the degree over $\mathfrak{k}'$ of the division algebra component of A divides $(\mathfrak{K} : \mathfrak{k}')^2$.

Since $(\mathfrak{K} : \mathfrak{k}')$ is a power of p, the preceding theorem shows that the exponent of A is a power of p. On the other hand, the exponent of A is a factor of the exponent of D, and is therefore prime to p. (This follows from a trivial generalization of Lemma 8.3F to several factors: $(D \times_{\mathfrak{k}} D \cdots \times_{\mathfrak{k}} D) \times_{\mathfrak{k}} \mathfrak{k}' = (D \times_{\mathfrak{k}} \mathfrak{k}') \times_{\mathfrak{k}'} (D \times_{\mathfrak{k}} \mathfrak{k}') \cdots \times_{\mathfrak{k}'} (D \times_{\mathfrak{k}} \mathfrak{k}')$, that is, $D^e \times_{\mathfrak{k}} \mathfrak{k}' = (D \times_{\mathfrak{k}} \mathfrak{k}')^e$.

This means that $A = D \times_{\mathfrak{k}} \mathfrak{k}'$ has exponent 1, so that $D \times_{\mathfrak{k}} \mathfrak{k}' \sim 1$ and $\mathfrak{k}'$ is a splitting field for D. But this is impossible; for if $\mathfrak{k}'$ were a splitting field for D, $(\mathfrak{k}' : \mathfrak{k})$ would by 8.3C be a multiple of $\sqrt{(D : \mathfrak{k})}$, and thus would be divisible by p, whereas $(\mathfrak{k}' : \mathfrak{k})$ is by construction prime to p.

CHAPTER IX

NONSEMISIMPLE RINGS WITH MINIMUM CONDITION

1. INTRODUCTION

In the previous chapters the study of semisimple rings was brought to that of sfields. Considerable progress has been made recently in the study of nonsemisimple rings R, that is, rings with radicals, but the theory is still incomplete. In particular, the theory of vector spaces over such rings R or, in other words, the representation theory of these rings, is only partially developed. In this chapter we aim to bring together some of the initial theorems for a theory of nonsemisimple rings with minimum condition. We shall confine the discussion to nonnilpotent rings and, more particularly, to rings with unit element, assuming throughout that our rings have the minimum condition. We shall also employ our previous convention of excluding the zero idempotent, so that statements concerning idempotents always refer to nonzero idempotents.

2. INDECOMPOSABLE SPACES OVER A RING

In the case of semisimple rings we have seen that an R-left space V may be decomposed into a direct sum of irreducible R-left spaces, and that an irreducible R-left space was characterized by its inverse R-homomorphism ring being a sfield. If R has a radical N, then for an R-left space V a decomposition into irreducible R-left spaces may no longer exist, and this leads us to consider <u>indecomposable</u> R-left spaces. An R-left space V is called indecomposable if V may not be written as a direct sum $V = V_1 + V_2$ of R-left spaces. An indecomposable R-left space V may contain an R-left subspace, in which case V would not be irreducible. In the characterization of indecomposable R-left spaces we need the concept of a <u>completely primary</u> ring. A ring R with radical N is called completely primary if R-N is a sfield and R has a unit element. Then the main theorem of indecomposable spaces is

Theorem 9.2A. *Let V be an R-left space such that the inverse R-homomorphism ring R' of V has minimum condition on left ideals. Then V is indecomposable if and only if R' is completely primary.*

Before proving the main theorem, we show

Theorem 9.2B. *Let R be a ring with minimum condition on left ideals, and with unit element e. Then R is completely primary if and only if e is the only idempotent contained in R.*

Proof. Assume e is the only idempotent contained in R. If α is any element of R which is not contained in the radical N of R, then the ideal $R\alpha$ is nonnilpotent; and, since the minimum condition holds in R, there exists an idempotent in $R\alpha$ which by assumption must be e. This gives that $R\alpha = R$, and so there exists an element β in R such that $\beta\alpha = e$. β is not nilpotent, for if $\beta^{\rho} = 0$, then $0 = \beta^{\rho}\alpha^{\rho-1} = \beta$. It follows that the elements of R which do not lie in N form a group under multiplication, and R-N is a sfield.

On the other hand, assume that R is completely primary. If R contains an idempotent $e_1 \neq e$, then e_1, $e-e_1$ would be mutually orthogonal idempotents, and their residue classes, modulo N, would be divisors of zero in R-N, contrary to the hypothesis that R-N is a sfield.

We shall now prove the main Theorem 9.2A.

If part. Let R' be completely primary, and suppose a decomposition $V = V_1 + V_2$ is possible. Denote by $e_1'(e_2')$ the element of R' which maps $V_1(V_2)$ identically into itself, and $V_2(V_1)$ into 0. Then e_1', e_2' would be mutually orthogonal idempotents in R', and by 9.2D there is only one nonzero idempotent in R'.

Only if part. Let V be an indecomposable R-space and suppose R' contains a nonzero idempotent e_1' which is different from e', the unit element of R'. Then any vector X of V may be written $Xe_1' + X(e' - e_1')$, and this would lead to a decomposition of V into $Ve_1' + V(e' - e_1')$, contradictory to our assumption that V is an indecomposable R-space. The only idempotent in R' is, accordingly, the unit element e', and by 9.2B R' is completely primary.

Let us now turn our attention to the ideals of R. We observe that, if $\mathfrak{l}$ is an indecomposable nonnilpotent left ideal of R, then, by Theorem 2.6B, there exists an idempotent e in $\mathfrak{l}$ such that $\mathfrak{l} = \mathfrak{l}e = Re$, no radical component appearing since $\mathfrak{l}$ is indecomposable. Again, by a familiar argument we can show that eRe is the inverse R-homomorphism ring of $\mathfrak{l} = Re$. Further, the minimum condition holds in eRe. For if $\mathfrak{m} \supset \mathfrak{m}_1 \supset \mathfrak{m}_2 \supset \cdots$ is a descending chain of left ideals of eRe, then $R\mathfrak{m} \supset R\mathfrak{m}_1 \supset R\mathfrak{m}_2 \supset \cdots$ is a descending chain of left ideals of R, and since this latter chain terminates, so must the former. Applying the main theorem, we obtain

Theorem 9.2C. A nonnilpotent left ideal $\mathfrak{l} = Re$ *is indecomposable if and only if the ring* eRe *is completely primary.*

Since the condition in 9.2C is two-sided, we have

Corollary 9.2D. If $\mathfrak{l} = Re$ *is an indecomposable nonnilpotent left ideal, then* $\mathfrak{r} = eR$ *is an indecomposable nonnilpotent right ideal.*

Another, and useful, characterization of indecomposable nonnilpotent ideals is given by

Theorem 9.2E. A nonnilpotent left ideal $\mathfrak{l} = Re$ *is indecomposable if and only if every subideal of* $\mathfrak{l}$ *is nilpotent.*

Proof. If every subideal of $\mathfrak{l}$ is nilpotent, then, if $\mathfrak{l} = \mathfrak{l}_1 + \mathfrak{l}_2$, we would have $\mathfrak{l}_1$, $\mathfrak{l}_2$ nilpotent and consequently also $\mathfrak{l}$ nilpotent, contrary to our assumption on $\mathfrak{l}$.

If $\mathfrak{l} = Re$ is indecomposable, then, by 9.2C, eRe is completely primary, and consequently, by 9.2B, e is the only idempotent in eRe. Let now $\mathfrak{l}_1$ be a nonnilpotent subideal of $\mathfrak{l}$, and e_1 an idempotent of $\mathfrak{l}_1$. Then $ee_1 \varepsilon eRe$ and is different from zero since $e_1^2 = e_1 e e_1$. Then $ee_1 = e$. It follows that $R(ee_1) = Re \subseteq \mathfrak{l}_1$, and so $Re = \mathfrak{l}_1$. This shows that the only nonnilpotent left ideal in Re is Re itself.

Remark. From 9.2E it appears that the terms "indecomposable" and "minimal" are interchangeable for nonnilpotent ideals--the indecomposable nonnilpotent ideals are the minimal nonnilpotent ideals, and conversely. Below we shall use the more customary term "indecomposable."

An important consequence of Theorem 9.2E is

Corollary 9.2F. Let $\mathfrak{l} = Re$ *be an indecomposable nonnilpotent left ideal of* R. *Then* Ne *is the unique maximal subideal of* $\mathfrak{l}$, *and* Re/Ne *is an irreducible* R-*left space.*

Proof. Every subideal of $\mathfrak{l}$ is nilpotent, and hence in the radical N of R, that is, every subideal is contained in $N \cap \mathfrak{l} = Ne$, so that Ne is the single maximal subideal of $\mathfrak{l}$. Since Ne is maximal, Re/Ne is irreducible.

Theorem 9.2G. Let $\mathfrak{l} = Re$, $\mathfrak{l}' = Re'$ *be two indecomposable nonnilpotent left ideals of* R. *Then* $\mathfrak{l}$ *is isomorphic to* $\mathfrak{l}'$ *if and only if* $V = Re/Ne$ *is isomorphic to* $V' = Re'/Ne'$.

Proof. Let $V \cong V'$. The residue class of e, modulo Ne, is a vector of V which is not annihilated by e. Since $V' \cong V$, there exists an element λ' in $\mathfrak{l}'$ such that $e\lambda' \neq 0 \pmod{Ne'}$. For convenience, we replace λ' by $e\lambda'$, and now have $e\lambda' = \lambda'$. Since $\lambda' \notin Ne'$, then $\lambda' \notin N$. Then the left ideal $\mathfrak{l}\lambda'$ is not in N, as $\mathfrak{l}\lambda'$ contains $e\lambda' = \lambda'$. Hence $\mathfrak{l}\lambda'$ is a nonnilpotent subideal of $\mathfrak{l}'$, and by Theorem 9.2E we obtain

$$\mathfrak{l}\lambda' = \mathfrak{l}'.$$

This maps $\mathfrak{l}$ into $\mathfrak{l}'$, but we are not yet sure that the mapping is (1-1). However, by similar reasoning we can find $\lambda = e'\lambda$ in $\mathfrak{l}$ such that

$$\mathfrak{l}'\lambda = \mathfrak{l}.$$

Since $e\lambda' = \lambda'$ and $\lambda \in Re$, we have $\lambda'\lambda$ is in eRe. Further, $\lambda'\lambda$ is not in N, for if it were, $\mathfrak{l} = \mathfrak{l}\lambda'\lambda$ would be nilpotent. By Theorem 9.2C, eRe is completely primary, so that $\lambda'\lambda$ has an inverse. Let $(\lambda'\lambda)\mu = e$. Then the mapping

$$\alpha \rightarrow \alpha' = \alpha\lambda'$$

of $\mathfrak{l}$ on $\mathfrak{l}'$ is (1-1), for if $\alpha \rightarrow 0$, that is, $\alpha\lambda' = 0$, then $\alpha\lambda'\lambda\mu = \alpha e = \alpha = 0$.

The converse statement, namely, that, if $\mathfrak{l} = Re \cong \mathfrak{l}' \cong Re'$, then $Re/Ne \cong Re'/Ne'$, is easy to prove. For if in the isomorphic mapping of $\mathfrak{l}$ on $\mathfrak{l}'$, $e \rightarrow \lambda'$, then $\alpha e \rightarrow \alpha\lambda'$, $\alpha \in R$, so that $R\lambda' = Re'$. Then λ' is not in N. Further, the mapping of $\mathfrak{l}$ on $\mathfrak{l}'$ induces a mapping of Re on V' in which

$$e \rightarrow [\lambda'] = \lambda' + Ne', \quad \alpha e \rightarrow [\alpha\lambda'].$$

The kernel of the mapping is Ne, and as V' is irreducible, we obtain $Re/Ne \cong V'$.

Theorem 9.2G will be applied below to set up a 1-1 correspondence between the distinct irreducible R-spaces and the distinct indecomposable nonnilpotent left ideals of R. We shall first proceed to examine the decomposition of R into indecomposable left ideals.

3. DECOMPOSITION OF A RING

In this section we shall study the decomposition* of R by comparison with the decomposition of R - N. An immediate remark is

Theorem 9.3A. R - N is a semisimple ring.

Proof. We must verify that $\bar{R} = R - N$ has no radical, and has the minimum condition on left ideals. Let $[\alpha]$ denote the residue class of the element α modulo N. Then we call the mapping $\alpha \rightarrow [\alpha]$ the natural homomorphism of R into $\bar{R}$. Let now $\bar{M}$ be a nilpotent left ideal of $\bar{R}$, and M denote the set of elements of R which map into $\bar{M}$ under the natural homomorphism. We see that M is a left ideal of R, and, further, M is nilpotent and contains N. Then M = N, so that $\bar{M}$ is the zero ideal of $\bar{R}$. Again, by use of the natural homomorphism we may see that any descending chain of ideals in $\bar{R}$ must terminate, since the descending chain of corresponding ideals in R terminates. Accordingly, the minimum condition holds in $\bar{R}$, and $\bar{R}$ is semisimple.

We shall require the following lemma.

Lemma 9.3B. The radical of the ring eRe, e *idempotent, is* eNe.

Proof. Let N' denote the radical of the ring eRe. Since eNe is a two-sided nilpotent ideal of eRe, then $eNe \subseteq N'$. On the other hand, if $\lambda \varepsilon N'$, $\alpha \varepsilon R$, then $(\alpha\lambda)^{n+1} = \alpha\lambda(e\alpha e\lambda)^n$, and, since $e\alpha e\lambda$ is nilpotent, it follows that $\alpha\lambda$ is nilpotent. This gives that $\lambda \varepsilon N$, and as $\lambda = e\lambda e$, then $\lambda \varepsilon eNe$, and $N' \subseteq eNe$, $N' = eNe$.

Theorem 9.3C. A decomposition of $\bar{R} = R - N$ *into minimal left ideals*

$$\bar{R} = \bar{R}_1\bar{e}_1 + \cdots \bar{R}\bar{e}_n, \tag{3.1}$$

where the $\bar{e}$ *are mutually orthogonal idempotents, leads to a decomposition*

$$R = Re_1 + \cdots Re_n + N_0, \tag{3.2}$$

with mutually orthogonal idempotents e_i, *where* e_i *lies in the residue class* $\bar{e}_i$, Re_i *is indecomposable, and* N_0 *is a nilpotent left ideal. Conversely, a decomposition* (3.2) *of* R *implies a decomposition* (3.1) *of* $\bar{R}$.

*For an alternative development of the decomposition of R see R. Brauer, "On the nilpotency of the radical of a ring," Bull. Amer. Math. Soc., 48 (1942), 752-758.

Proof. We assume that mutually orthogonal idempotents $e_1, \cdots, e_{k-1}$ have been obtained from the classes $\bar{e}_1, \cdots, \bar{e}_{k-1}$, respectively, and seek a suitable idempotent e_k from the class $\bar{e}_k$. Let λ be an element of $\bar{e}_k$, and take

$$(3.3)\quad \lambda_1 = \lambda - \lambda(e_1 + \cdots + e_{k-1}) - (e_1 + \cdots + e_{k-1})\lambda + (e_1 + \cdots + e_{k-1})\lambda(e_1 + \cdots + e_{k-1}).$$

Since $e_i\lambda \equiv \lambda e_i \equiv 0 \pmod{N}$, we have $\lambda_1 \equiv \lambda \pmod{N}$, and so λ_1 lies in the class $\bar{e}_k$. Then $x_1 = \lambda_1^2 - \lambda_1 \equiv 0 \pmod{N}$. If λ_1 is not idempotent, we take $\lambda_2 = \lambda_1 + x_1 - 2\lambda_1 x_1$ (cf. proof of Theorem 2.4A), then $\lambda_2^2 - \lambda_2 = 4x_1^3 - 3x_1^2$, and proceeding in this way we obtain a λ_r with $\lambda_r^2 - \lambda_r = 0$. We observe from (3.3) that $e_i\lambda_1 = \lambda_1 e_i = 0$ for $i < k$, and then $e_i\lambda_r - \lambda_r e_i = 0$, since λ_r may be written as a polynomial in λ_1. We choose $e_k = \lambda_r$. In this manner we obtain n mutually orthogonal idempotents $e_1, \cdots, e_n$. We take $e = e_1 + \cdots + e_n$ and form the Peirce decomposition $\alpha = \alpha e + (\alpha - \alpha e)$, $\alpha \varepsilon R$. Since $\bar{e} = \bar{e}_1 + \cdots + \bar{e}_n$ is the unit of $\bar{R}$, the element $\alpha - \alpha e \equiv 0 \pmod{N}$; hence the elements $\alpha - \alpha e$ give a nilpotent left ideal N_0 of R.

The inverse R-homomorphism ring e_iRe_i of Re_i is completely primary. For from Lemma 9.3B, e_iNe_i is the radical of e_iRe_i, and $e_iRe_i - e_iNe_i \cong \overline{e_iRe_i} = \bar{e}_i\bar{R}\bar{e}_i$, which is a sfield since $\bar{R}\bar{e}_i$ is a minimal left ideal of $\bar{R}$. From Theorem 9.2A it follows that Re_i is indecomposable, and this completes the first part of the theorem.

For the converse we observe that, if Re_i is indecomposable, then e_iRe_i is completely primary, and $\bar{e}_i\bar{R}\bar{e}_i$ is a sfield, so that $\bar{R}\bar{e}_i$ is a minimal left ideal (cf. Theorem 5.4A). Further, the $\bar{e}_i (i = 1, 2, \cdots, n)$ form a set of mutually orthogonal idempotents of $\bar{R}$.

Remark. If R possesses a unit element e', then from the decomposition (3.2) we have

$$e' = \rho_1 e_1 + \cdots + \rho_n e_n + (e' - e'e) = \rho_1 e_1 + \cdots + \rho_n e_n.$$

It follows that $e'e = e'(e_1 + \cdots + e_n) = e' = e$. Then, for a ring R with unit element e, the decomposition (3.2) becomes

$$(3.3)\quad R = Re_1 + \cdots + Re_n, \quad e = e_1 + e_2 + \cdots + e_n,$$

since now the elements $\alpha - \alpha e$ of N_0 are all zero. Applying Corollary 9.2D, we obtain a decomposition

$$(3.4)\quad R = e_1R + \cdots + e_nR$$

of R into indecomposable right ideals.

In what follows we shall always assume that our rings R possess unit elements.

Theorem 9.3D. *If $R = Re_1 + \cdots + Re_n$ is a decomposition of R into indecomposable nonnilpotent left ideals, then any irreducible R-space V such that $RV \neq 0$ is isomorphic to one of the spaces Re_i/Ne_i $(i = 1, 2, \cdots, n)$.*

Proof. It is impossible that all $e_1, \cdots, e_n$ annihilate V, since $RV \neq 0$. Suppose, then, that $Re_iV \neq 0$, and that X is a vector of V such that $Re_iX \neq 0$. We obtain that $Re_iX = V$, since V is irreducible. The elements of Re_i which annihilate X form a left ideal $\mathfrak{m}$, and we have $V \cong Re_i/\mathfrak{m}$. But, since V is irreducible, $\mathfrak{m}$ must be maximal, so that, by Corollary 9.2F, $\mathfrak{m} = Ne_i$, and the theorem is proved.

Isomorphism is an equivalence relation by means of which we may classify into disjoint classes the irreducible R-left spaces and also the indecomposable nonnilpotent R-left ideals, Re. Then we obtain

Theorem 9.3E. *There is a one-to-one correspondence between the classes of irreducible R-left spaces and the classes of indecomposable nonnilpotent left ideals Re.*

Proof. For, by Theorem 9.3D, any irreducible R-left space with $RV \neq 0$ is isomorphic to one of the spaces Re_i/Ne_i, and by Theorem 9.3B two irreducible spaces Re_i/Ne_i, Re_j/Ne_j are isomorphic if and only if the corresponding indecomposable left ideals Re_i, Re_j are isomorphic.

4. LOEWY SERIES

We saw in Theorem 5.3H that, if R is a semisimple ring and V is an R-space with minimum condition such that $RV = V$, then V is the direct sum of a finite number of irreducible spaces. For the case of rings R with radical, one of the main concepts is that of Loewy series. Let V be an R-left space. A series of subspaces of V

$$(4.1) \qquad V \supset V_1 \supset V_2 \supset \cdots \supset V_{r-1} \supset (0)$$

is called a Loewy series for V if each space V_{i-1}/V_i $(i = 1, 2, \cdots, r)$ is a direct sum of irreducible R-spaces. We shall assume always that the minimum condition holds in V, then it follows that each space V_{i-1}/V_i is a sum of a finite number of irreducible spaces. We also assume that for spaces V with operator ring $\overline{R} = R - N$ the condition $\overline{R}V = V$ is satisfied.

It is convenient here to introduce the usual term completely reducible—an R-space V being called completely reducible if V is a direct sum of irreducible R-spaces.

As a starting point for the study of the special Loewy series that we shall discuss, we take the following theorem.

Theorem 9.4A. Let W be an R-left space, and W_1 a subspace of W. Then the R-left space W/W_1 is completely reducible if and only if $NW \subseteq W_1$.

Proof. If $NW \subseteq W_1$, then W/W_1 as an R-space behaves the same as W/W_1 as an R-N space. For if $\alpha \equiv \alpha_1$, mod N, $\alpha = \alpha_1 + \eta$, $\eta \varepsilon$ N, then for any vector X of W, $\alpha X = \alpha_1 X + \eta X \equiv \alpha_1 X \pmod{W_1}$, that is, α, α_1 produce the same operation on W/W_1. But W/W_1 as an R - N space is completely reducible, since R - N is semisimple.

For the second part of the theorem, suppose W/W_1 is completely reducible

$$W/W_1 = V_1 + V_2 + \cdots + V_t,$$

where the V_i are irreducible, and assume $NW \not\subseteq W_1$. Then $NV_i \neq 0$ for some V_i. The argument is now classical. V_i is irreducible, and NV_i is a nonzero subspace, hence $NV_i = V_i$. But then $N^\rho V_i = V_i$ for any positive integer ρ, and for sufficiently large ρ, $N^\rho V = 0$, which gives a contradiction.

The particular series for an R-space V that we are interested in are the upper Loewy series for V and its dual, the lower Loewy series. The upper series is formed by choosing in V such a space V_1 that V/V_1 is a maximal space in the set of completely reducible R-spaces, then a subspace V_2, so that V_1/V_2 is maximal in the set of completely reducible R-spaces, and so on. It follows from this definition that the upper Loewy series is unique. For, if V_1 and V_1' are two subspaces of V such that V/V_1, V/V_1' are maximal completely reducible spaces, then $V/V_1 \cap V_1'$ is also completely reducible, since, by Theorem 9.4A, $NV \subseteq V_1$, $NV \subseteq V_1'$, and hence $NV \subseteq V_1 \cap V_1'$. But $V/V_1 \cap V_1' \supseteq V/V_1$, and as V/V_1 is maximal, the equality must hold, and this implies $V_1 \subseteq V_1'$. Similarly, $V_1' \subseteq V_1$, that is, $V_1 = V_1'$. In fact, we may observe that V_1 is NV, the intersection of all subspaces V' of V such that V/V' is completely reducible.

The lower Loewy series is formed by taking the subspace W_1 of V, which is maximal, as a completely reducible space, then W_2, so that W_2/W_1 is maximal as a completely reducible space, and so on, until V is reached. It follows that the lower Loewy series is unique, in fact, W_i is the join of all subspaces W' of V such that W'/W_{i-1} is completely reducible.

Let $N^{m-1}V \neq (0)$, $N^m V = (0)$. Then we have

Theorem 9.4B. *The series*

$$V \supset NV \supset N^2V \supset \cdots \supset N^{m-1}V \supset N^mV \supset (0)$$

is the upper Loewy series of V,

and

Theorem 9.4C. *Let* $M_\rho(V)$ *denote the subspace of* V *which is annihilated by* N^ρ *but not by* $N^{\rho-1}$. *In particular,* $V = M_m(V)$. *Then*

$$V = M_m(V) \supset M_{m-1}(V) \supset \cdots \supset M_1(V) \supset (0)$$

is the lower Loewy series of V. *The upper and the lower Loewy series have the same length* m.

By using Theorem 9.4A these statements above follow at once from the definitions of the upper and the lower Loewy series.

Corollary 9.4D. $Re \supset NE \quad \cdots \supset N^{n-1}e \supset (0)$ *is the upper Loewy series of* Re.

Theorem 9.4E. *The* τ-*th completely reducible space* $N^{\tau-1}V/N^\tau V$ *of the upper Loewy series has at least one irreducible constituent in common with the* τ-*th completely reducible space* $M_{m-\tau+1}(V)/M_{m-\tau}(V)$ *of the lower series*.

Proof. From definition we have that, for $\rho \leq m$, $N^\rho V \subseteq M_{m-\rho}(V)$. Further, $N^\rho V \not\subseteq M_{m-\rho-1}(V)$, for the contrary implies that $N^{m-\rho-1}(N^\rho V) = 0$, that is, $N^{m-1}V = 0$. Then $N^\tau V$ is contained in both $N^{\tau-1}V$ and $M_{m-\tau}(V)$, so that $N^\tau V \subseteq N^{\tau-1}V \cap M_{m-\tau}(V)$. $M_{m-\tau}(V)$ is contained as a proper space in $M_{m-\tau}(V) + N^{\tau-1}V$, which, in turn, is a subspace of $M_{m-\tau+1}(V)$. From the relation

$$(M_{m-\tau}(V) + N^{\tau-1}V)/M_{m-\tau}(V) \cong N^{\tau-1}V/N^{\tau-1}V \cap M_{m-\tau}(V)$$

it follows, then, that $M_{m-\tau+1}(V)/M_{m-\tau}(V)$ has a subspace in common with $N^{\tau-1}V/N^\tau V$, which completes the proof of the theorem.

5. CARTAN INVARIANTS

Let $\mathfrak{l} = Re$ be an indecomposable ideal of R, and V any R-left space with finite composition series:

(5.1) $$V = V_0 \supset V_1 \supset V_2 \supset \cdots \supset V_{r-1} \supset V_r = (0).$$

Here each V_i/V_{i+1} is simple, that is, is an irreducible R-space. We shall refer to the V_i/V_{i+1} as the <u>composition spaces</u> of V. Our question is, What are the homomorphic mappings of $\mathfrak{l} = Re$ into V?

Let σ be a homomorphic mapping of $\mathfrak{l}$ into V, and let $\sigma(e) = X_0$. Then $\alpha \cdot e \to \alpha \cdot X_0$, so that $\sigma(\alpha) = \alpha X_0$. We observe that $\sigma(e) = eX_0 = X_0$, so that $X_0 \in eV$. Conversely, every element of eV gives a homomorphic mapping of $\mathfrak{l}$ into V, namely, if X is any element of eV, then $eX = X$, and $\alpha \to \alpha X$ is a homomorphism of Re into V. We consider two cases for each irreducible space V_i/V_{i+1}.

(1) $Re/Ne \cong V_i/V_{i+1}$. Then a vector $[X_i]$ in V_i/V_{i+1} is not annihilated by e, that is, $eX_i \notin V_{i+1}$. For convenience we replace X_i by eX_i, and we then have a vector X_i in V_i such that

$$X_i = eX_i \notin V_{i+1},$$

hence

$$X_i \notin eV_{i+1}.$$

Then ReX_i, which contains eX_i, is a subspace of V_i, which is not in V_{i+1}, and then, since V_i/V_{i+1} is irreducible, we have

(5.2) $$ReX_i = V_i \pmod{V_{i+1}},$$

and

(5.3) $$eReX_i = eV_i \pmod{eV_{i+1}}.$$

(2) Re/Ne is not isomorphic to V_i/V_{i+1}. Then $eV_i = 0 \pmod{V_{i+1}}$. For if X, an element of eV_i, were such that $eX \neq 0 \pmod{V_{i+1}}$, then the irreducible space V_i/V_{i+1} would be isomorphic to Re/Ne. From $eV_i \subseteq V_{i+1}$ we have $eV_i \subseteq eV_{i+1}$, and so $eV_i = eV_{i+1}$.

Looking over the whole chain (5.1), we select vectors $X_{i_1}, \cdots, X_{i_s}$ for the cases where Re/Ne is isomorphic to the composition spaces V_i/V_{i+1}. By the above argument, $eRe\, X_{i_\rho} = eV_{i_\rho} \pmod{eV_{i_\rho+1}}$, while $eV_{j_\rho} = eV_{j_\rho+1}$ in the nonisomorphic cases. Hence,

(5.4) $$eV = eRe\, X_{i_1} + \cdots + eRe\, X_{i_s}.$$

<u>Theorem 9.5A. Let V be an R-left space and</u>

$$V \supset V_1 \supset V_2 \supset \cdots \supset V_{r-1} \supset V_r = (0)$$

<u>be a composition series of V. Every element of the additive</u>

group eV defines a homomorphic mapping of the indecomposable left ideal Re into V. Moreover, if s denotes the number of composition spaces of V which are isomorphic to Re/Ne, then there exist s vectors $X_{i_\rho} \subset eV$ such that

$$eV = eRe\, X_{i_1} + \cdots + eRe\, X_{i_s}.$$

Remark. From (5.4) it follows that s is the composition length of eV considered as an eRe space. If the completely primary ring eRe may be expressed as

$$eRe = \mathfrak{k} + eNe, \tag{5.5}$$

where $\mathfrak{k}$ is a sfield isomorphic to eRe - eNe, then we may show that

$$eV = \mathfrak{k}X_{i_1} + \cdots + \mathfrak{k}X_{i_s}. \tag{5.6}$$

For let X_i be chosen as in case (1) above, then NeX_i (mod V_{i+1}) is a subspace of V_i (mod V_{i+1}), and as V_i/V_{i+1} is irreducible, then $NeX_i = 0$ (mod V_{i+1}), and so $eNeX_i = 0$ (mod eV_{i+1}). Then, where (5.5) holds, we may replace (5.3) by

$$\mathfrak{k}\, X_i = eV_i \pmod{eV_{i+1}}. \tag{5.7}$$

In this case we have from (5.6) that s is the $\mathfrak{k}$-dimension of the space eV.

We suppose R has the decomposition $R = Re_1 + \cdots Re_n$ into indecomposable left ideals, and apply Theorem 9.5A to the completely reducible spaces $N^\tau e_j/N^{\tau+1}e_j$ of the upper Loewy series for Re_j.

Corollary 9.5B. The number of irreducible subspaces of the completely reducible space $N^\tau e_j/N^{\tau+1}e_j$ which are isomorphic to the irreducible space Re_i/Ne_i is equal to the composition length t^τ_{ij} of $e_iN^\tau e_j/e_iN^{\tau+1}e_j$ as an e_iRe_i space.

Corollary 9.5C. The number of composition spaces of Re_j which are isomorphic to Re_i/Ne_i is equal to the composition length $\tilde{c}_{ij}$ of e_iRe_j as an e_iRe_i space. $\tilde{c}_{ij} = \Sigma^{m-1}_{\tau=0}t^\tau_{ij}$, where m is the length of the Loewy series of Re_j.

We observe that if $Re_{i'} \cong Re_i$, and $Re_{j'} \cong Re_j$, then $\tilde{c}_{i'j'} = \tilde{c}_{ij}$. Let $A_1, A_2, \cdots, A_k$ denote the classes of isomorphic indecomposable nonnilpotent left ideals Re, and let c_{ij} denote the value of $\tilde{c}_{i_\alpha j_\beta}$ for any Re_{i_α} belonging to A_i, and any Re_{j_β} belonging to A_j. These numbers c_{ij}

have been called the Cartan invariants,* and are fundamental in the theory of nonsemisimple systems. Corollary 9.5C provides us with two characterizations of these invariants.

6. BLOCKS AND TWO-SIDED IDEALS

We here consider the decomposition of R into two-sided ideals. Let R have the decomposition $R = Re_1 + \cdots + Re_n$ into indecomposable left ideals. We have seen (cf. Corollary 9.5C) that a necessary and sufficient condition that Re_j should have a composition space which is isomorphic to Re_i/Ne_i is that $e_iRe_j \neq 0$. We now say that e_i and e_j belong to the same <u>block</u>, if there exists a chain

$$e_i, e_r, \cdots, e_s, e_j,$$

such that any two neighboring idempotents e_g, e_h in the chain define left ideals Re_g, Re_h, which have at least one composition space in common. This is evidently a reflexive, symmetric, and transitive relation, by means of which the idempotents are classified into disjoint classes of equivalent elements. Let

$$B_1, B_2, \cdots, B_t$$

denote the totality of blocks. We denote by R_ρ the direct sum of the ideals Re_i with $e_i \in B_\rho$. Then we have

<u>Theorem 9.6A</u>. $R = R_1 + R_2 + \cdots + R_t$ <u>is the unique decomposition of R into a direct sum of indecomposable two-sided ideals</u>.

<u>Proof</u>. R_ρ is a left ideal; it remains to be shown that R_ρ is also a right ideal. We observe, first, that, if e_i, e_j belong to different blocks, then $Re_i \cdot Re_j = 0$. For, if $Re_i \cdot Re_j \neq 0$, then $e_iRe_j \neq 0$, so that Re_j has a composition space which is isomorphic to Re_i/Ne_i, and then e_i, e_j would be a proper chain joining e_i and e_j, contrary to our assumption that e_i and e_j belong to different blocks. It follows now that

$$R_\rho R = R_\rho(Re_1 + \cdots + Re_n) \subseteq R_\rho \cdot R_\rho \subseteq R_\rho,$$

that is, R_ρ is also a right ideal.

We shall now show that the decomposition is unique. Let $R = R_1' + R_2' + \cdots + R_s'$ be any decomposition of R into

*See the note by R. Brauer and C. Nesbitt, "On the regular representations of algebra," <u>Proc. Nat. Acad. Sci.</u>, Vol. 23 (1937).

a direct sum of indecomposable two-sided ideals. Then one of the indecomposable left ideals Re_i may be contained in only one summand R'_ρ. For $Re_i = R'_1e_i + R'_2e_i + \cdots + R'_se_i$, and, if more than one $R'_\rho e_i$ were different from 0, we would have a decomposition of Re_i. Then Re_iR is contained in just one summand R'_ρ. Further, if $e_iRe_j \neq 0$, then since e_iRe_j is contained in both Re_iR and Re_jR, the two-sided ideals Re_iR, Re_jR must lie in the same summand R'_ρ.

It follows easily now that, if e_i, e_j lie in the same block, then Re_iR, Re_jR must be contained in the same summand R'. For let

$$e_i,\ e_r, \cdots,\ e_u,\ e_v, \cdots,\ e_s,\ e_j$$

be a chain connecting e_i e_j, and let the composition space common to Re_u, Re_v be isomorphic to Re_h/Ne_h. Then $e_hRe_u \neq 0$, and $e_hRe_v \neq 0$, so that Re_uR, Re_vR lie in the same summand R'_ρ of R as does Re_hR. Applying this argument all along the chain, we find that Re_iR, Re_jR are in the same summand. Thus each R'_ρ contains, and hence must coincide with, some R_σ of our decomposition of R into two-sided ideals R_ρ corresponding to the blocks B_ρ.

7. APPLICATIONS TO ALGEBRAS

We shall give here a few illustrative applications of the foregoing theorems to the theory of algebras. In the following argument A denotes an algebra over a field $\mathfrak{K}$. We assume that A has a unit element e.

Let B: $\alpha \rightarrow B_\alpha$, $\alpha \in A$ be a representation of A with coefficients in the field $\mathfrak{K}$, and let V be the corresponding representation space, V being considered as an A-left space. If we consider the A-homomorphisms σ of V also as left operators of V, we obtain a set of matrices C_σ which are commutative with the matrices B_α. The matrices C_σ give the commutator algebra C of B. Theorem 9.2A yields that the representation B is indecomposable if and only if the commutator algebra C of B is completely primary. We assume now that B is indecomposable, so that C is completely primary, and we examine the irreducible constituents of C. If N_c denotes the radical of C, then $\overline{C} = C - N_c$ is a sfield. Let $V = V_1 \supset V_2 \supset \cdots \supset V_r$ be a composition series of V considered as a C-left space. As N_c annihilates the irreducible spaces V_{i-1}/V_i, then C operating on V_{i-1}/V_i is equivalent to the sfield $\overline{C}$ operating on V_{i-1}/V_i. Then all the V_{i-1}/V_i are isomorphic; in fact, each is isomorphic to $\overline{C}$. That is, all the irreducible constituents of C are equivalent and are isomorphic to $\overline{C}$.

We turn now to the regular representations of A. Let $A = Ae_1 + \cdots + Ae_n$ be a decomposition of A into indecomposable nonnilpotent left ideals. Each summand Ae_i, if considered as an A-left space, determines an indecomposable representation S^i: $\alpha \to S^i(\alpha)$, of A. S^i appears as an indecomposable part of the regular representation S of A (see §3.2). The upper Loewy series for Ae_i is $Ae_i \supset Ne_i \supset N^2e_i \supset \cdots \supset N^{t-1}e_i$. Let $(\varepsilon_1, \varepsilon_2, \cdots, e_{s_i})$ be a $\mathfrak{K}$-basis for Ae_i adapted to this Loewy series. From $\alpha(\varepsilon_1, \cdots, \varepsilon_{s_i}) = (\varepsilon_1, \cdots, \varepsilon_{s_i}) S^i(\alpha)$ we find that

$$S^i = \left\| \begin{matrix} S_1^i & & & & \\ & S_2^i & & & \\ & & \cdot & & \\ & * & & \cdot & \\ & & & & \cdot \\ & & & & S_t^i \end{matrix} \right\|$$

Here the S_q^i are completely reducible, that is, S_q^i may be decomposed in the form

$$S_q^i = \left\| \begin{matrix} F_1^q & & & \\ & \cdot & & \\ & & \cdot & \\ & & & \cdot \\ & & & F_m^q \end{matrix} \right\|,$$

where the F_σ^q are irreducible. In fact, the S_q^i are the uniquely determined (up to equivalence) upper Loewy constituents of S^i.* S_1^i is the maximal completely reducible part that can be brought as an upper constituent in S^i. Applying Corollary 9.2F we have that Ae_i/Ne_i is irreducible, and so S_1^i is irreducible. From Theorem 9.2G, S^i is equivalent to S^j if and only if S_1^i is equivalent to S_1^j. Again, from Theorem 9.3E we obtain a 1-1 correspondence between the classes of equivalent indecomposable parts of the regular representation S of A and the classes of equivalent irreducible representations of A.

Similarly, each summand e_iA in the decomposition $A = e_1A + \cdots + e_nA$ determines an indecomposable part T^i of the regular representation T of A. The upper Loewy

---

*See R. Brauer, "On sets of matrices with coefficients in a division ring," Trans. Amer. Math. Soc., 49 (1941), 502-548.

series for e_iA is $e_iA \supset e_iN \supset \cdots$; choosing a coördinate system $\begin{bmatrix} \delta_1 \\ \vdots \\ \delta_{t_i} \end{bmatrix}$ for e_iA adapted to this series, we obtain from

$$\begin{bmatrix} \delta_1 \\ \vdots \\ \delta_{t_i} \end{bmatrix} \alpha = T^i(\alpha) \begin{bmatrix} \delta_1 \\ \vdots \\ \delta_{t_i} \end{bmatrix}$$

$$T^i = \begin{Vmatrix} T_1^i & & & * \\ & T_2^i & & \\ & & \ddots & \\ & & & T_u^i \end{Vmatrix}$$

Here T_1^i corresponds to the irreducible space e_iA/e_iN, and so is irreducible and equivalent to S_1^i.

We shall assume now that the field $\mathfrak{K}$ is perfect (vollkommen), so that the completely primary algebras $e_i Ae_i$ over $\mathfrak{K}$ split, $e_iAe_i = \mathfrak{k}_i + e_iNe_i$, $\mathfrak{k}_i$ a sfield isomorphic to $e_iAe_i - e_iNe_i$. Let r_i denote the rank of $\mathfrak{k}_i$ over $\mathfrak{K}$. From Corollary 9.5C we have that the number of irreducible constituents of S which are equivalent to S_1^i is equal to $\Sigma_{j=1}^n \tilde{c}_{ij}$. We consider now $e_iA = e_iAe = \Sigma_{j=1}^n e_iAe_j$. Here e_iAe_j has dimension $\tilde{c}_{ij}$ as a $\mathfrak{k}_i$-space; hence the dimension t_i of e_iA as a $\mathfrak{K}$-space is $r_i\Sigma_{j=1}^n\tilde{c}_{ij}$. This gives that the number of irreducible constituents of S which are equivalent to S_1^i is equal to $\frac{t_i}{r_i}$, where t_i is the degree of the indecomposable part T^i of the regular representation T. By deriving for right spaces the analogue of Corollary 9.5C one obtains by a similar computation that the number of irreducible constituents of T which are equivalent to T_1^i is equal to $\frac{s_i}{r_i}$, where s_i is the degree of the indecomposable part S^i of S. We may interpret r_i as the rank over $\mathfrak{K}$ of the commutator algebra of $S_1^i \cong T_1^i$.

On the other hand, the number h_i of indecomposable parts S_j of S which are equivalent to S_i is equal to the number of indecomposable left ideals Ae_j in the decomposition $Ae_1 + \cdots + Ae_n$ which are isomorphic to Ae_i, and this, in turn, equals the number of minimal left ideals $\bar{A}\bar{e}_j$ in the decomposition $\bar{A}\bar{e}_1 + \cdots \bar{A}\bar{e}_n$ of $\bar{A} = A - N$ which are isomorphic to $\bar{A}\bar{e}_i$. This last number is equal to the dimension of $\bar{A}\bar{e}_i$ as a $\mathfrak{k}_i$-space. $\bar{A}\bar{e}_i$, considered as an

A-space, is isomorphic to Ae_i/Ne_i. Hence Ae_i/Ne_i has dimension h_i as a $\mathfrak{t}_i$-space, and dimension h_ir_i as a $\mathfrak{K}$-space. Let f_i denote the degree of S_1^i. We have, then, that $f_i = h_ir_i$. This gives that the number h_i of indecomposable parts of S which are equivalent to S^i equals $\frac{f_i}{r_i}$. Similarly, the number of indecomposable parts of R which are equivalent to T^i is $\frac{f_i}{r_i}$.

A final remark is that we may classify the indecomposable parts S^i according to blocks: S^i, S^j belonging to the same block if there exists a chain

$$S^i,\ S^p, \cdots,\ S^q,\ S^j,$$

such that any two neighboring indecomposable parts in the chain have at least one irreducible constituent in common.

8. MODULAR THEORY

In this section we shall discuss a main theorem of modular representations which connects the Cartan invariants in the modular theory with certain decomposition numbers. A development of this theorem has been given by Brauer,* and we shall follow his method.

Let $\mathfrak{K}$ be an algebraic number field, and $\mathfrak{p}$ a prime ideal of the ring $\mathfrak{o}$ of integers of $\mathfrak{K}$. Let $\pi \equiv 0 \pmod{\mathfrak{p}}$ but $\pi \not\equiv 0 \pmod{\mathfrak{p}^2}$. Then every element a of the $\mathfrak{p}$-adic number field $\mathfrak{K}_\mathfrak{p}$ has the form

(8.1) $a = c_{-m}\pi^{-m} + c_{-m+1}\pi^{-m+1} + \cdots + c_0 + c_1\pi + c_2\pi^2 + \cdots,$

where the c_i are integers from $\mathfrak{o}$, and $c_{-m} \not\equiv 0 \pmod{\mathfrak{p}}$. For a $\mathfrak{p}$-adic integer of $\mathfrak{K}_\mathfrak{p}$ the series (8.1) begins with $c_\nu\pi^\nu$, $\nu \geqq 0$. We denote the ring of $\mathfrak{p}$-adic integers by $\mathfrak{J}_\mathfrak{p}$.

Let A be an algebra with unit element and of rank n over the field $\mathfrak{K}_\mathfrak{p}$. We define, in the sense used by Brauer, a domain of integrity $R \subseteq A$ by

(1) R is a subring of A;

(2) R contains n linearly independent elements of A;

*See R. Brauer, "On modular and p-adic representations of algebras," Proc. Nat. Acad. Sci., 25 (1939), 252-258. Our method is also closely related to that used by T. Nakayama in "Some remarks on regular representations, induced representations, and modular representations," Ann. of Math., 39 (1938), 361-369.

(3) the elements of R when expressed by a basis $\eta_1, \eta_2, \cdots, \eta_n$ of A have the form $\Sigma a_i \eta_i$ with $a_i = b_i/w$, where the $b_i \in \mathfrak{J}_{\mathfrak{p}}$ and $w \in \mathfrak{J}_{\mathfrak{p}}$ is a fixed denominator;

(4) R contains the ring $\mathfrak{J}_{\mathfrak{p}}$ of $\mathfrak{p}$-adic integers.

A subring R of A fulfilling (1) to (4) may be obtained in the following way from A. Let $\varepsilon_1, \varepsilon_2, \cdots, \varepsilon_n$ be a $\mathfrak{R}_{\mathfrak{p}}$-basis of A such that ε_1 is the unit element of A. Let $\varepsilon_i \varepsilon_j = \Sigma_{k=1}^{n} a_{ijk} \varepsilon_k$, and suppose s is the smallest integer such that $\pi^s a_{ijk} \in \mathfrak{J}_{\mathfrak{p}} (i, j, k = 1, 2, \cdots, n)$. Then the set of elements $\varepsilon_1, \pi^s \varepsilon_2, \cdots, \pi^s \varepsilon_n$ is an $\mathfrak{J}_{\mathfrak{p}}$-basis for a subring R of A with unit element

$$R = \mathfrak{J}_{\mathfrak{p}} \varepsilon_1 + \mathfrak{J}_{\mathfrak{p}} \pi^s \varepsilon_2 + \cdots + \mathfrak{J}_{\mathfrak{p}} \pi^s \varepsilon_n.$$

R is a domain of integrity of A in the sense above.

<u>Lemma 9.8A.</u> <u>Let</u> V <u>be a</u> $\mathfrak{R}_{\mathfrak{p}}$ <u>space of dimension</u> r <u>and</u> C <u>be an</u> $\mathfrak{J}_{\mathfrak{p}}$ <u>subspace of</u> V <u>with a finite number of generators</u> $C = \mathfrak{J}_{\mathfrak{p}} X_1 + \cdots + \mathfrak{J}_{\mathfrak{p}} X_s$, $X_i \in V$, $i = 1, 2, \cdots, s$. <u>Then if</u> $D \subseteq C$ <u>is an</u> $\mathfrak{J}_{\mathfrak{p}}$ <u>subspace of</u> C, <u>there exist linearly independent vectors</u> $Y_1, \cdots, Y_p$ <u>in</u> D, $p \leq r$, <u>which form an</u> $\mathfrak{J}_{\mathfrak{p}}$ <u>basis for</u> D, <u>that is,</u> $D = \mathfrak{J}_{\mathfrak{p}} Y_1 + \cdots + \mathfrak{J}_{\mathfrak{p}} Y_p$ <u>(direct sum).</u>

<u>Proof.</u> Let $Z_1, \cdots, Z_r$ be a $\mathfrak{R}_{\mathfrak{p}}$ basis for V, and with regard to this basis let the components of X_i be $(a_{i1}, \cdots, a_{ir})$. We shall employ the exponential valuation of the elements of $\mathfrak{R}_{\mathfrak{p}}$ determined by assigning to each element a of $\mathfrak{R}_{\mathfrak{p}}$ the exponent of the leading term appearing in the expression (8.1) for a. As the a_{ij} $(i = 1, 2, \cdots, s; j = 1, 2, \cdots, r)$ form a finite set, their values have a lower bound, say m. Now let T be an element of D, $T = c_1^T X_1 + \cdots + c_s^T X_s$. With regard to the basis $Z_{\mathfrak{R}}$ of V, T has the unique set of components $(c_\nu^T a_{\nu 1}, \cdots, c_\nu^T a_{\nu r})$, where the repeated ν denotes summation over the range $1, \cdots, s$. Since the c_ν^T are integers from $\mathfrak{J}_{\mathfrak{p}}$, m is a lower bound for the values of the components $c_\nu^T a_{\nu k}$. Consider now the first components $c_\nu^T a_{\nu 1}$ of all vectors T in D, and assume that for some T, $c_\nu^T a_{\nu 1} \neq 0$. (If all $c_\nu^T a_{\nu 1} = 0$, we proceed at once to the second components.) We select from among all the vectors in D a vector Y_1 such that $c_\nu^{Y_1} a_{\nu 1}$ has lowest value. Then for each $T \in D$ there exists a unique integer $b_1^T \in \mathfrak{J}_{\mathfrak{p}}$ such that $T - b_1^T Y_1$ has components of the form $(0, d_2^T, \cdots, d_r^T)$. If for some T, $d_2^T \neq 0$, we select from all the vectors $T - b_1^T Y_1$ a vector Y_2 such that $b_2^{Y_2}$ has lowest value; if all b_2^Y are zero, we pass to the

next component. In this way we obtain independent vectors $Y_1, Y_2, \cdots, Y_p$, $p \leqq r$, which form an $\mathfrak{J}_\mathfrak{p}$-basis for D.

Corollary 9.8B. _If R is a domain of integrity in A, then R contains an $\mathfrak{J}_\mathfrak{p}$-basis._

Proof. By (3) of the definition of a domain of integrity, $R \subseteq \mathfrak{J}_\mathfrak{p}(\frac{\eta_1}{w}) + \cdots + \mathfrak{J}_\mathfrak{p}(\frac{\eta_n}{w})$. By (2) of the definition, R contains n linearly independent elements of A. Then, from 9.8A, with D = R, $C = \mathfrak{J}_\mathfrak{p}(\frac{\eta_1}{w}) + \cdots + \mathfrak{J}_\mathfrak{p}(\frac{\eta_n}{w})$, it follows that there exist n elements $\zeta_1, \cdots, \zeta_n$ of R such that $R = \mathfrak{J}_\mathfrak{p}\zeta_1 + \cdots + \mathfrak{J}_\mathfrak{p}\zeta_n$.

Lemma 9.8C. _Let V be an A-space, and R a domain of integrity in A. There exists an R-subspace W of V which spans V and has an $\mathfrak{J}_\mathfrak{p}$-basis._

Proof. Let $X_1, \cdots, X_f$ be a $\mathfrak{K}_\mathfrak{p}$-basis for V, and let $\zeta_1, \cdots, \zeta_n$ be an $\mathfrak{J}_\mathfrak{p}$-basis for R, as in Corollary 9.8B. Then $W = \Sigma_{i,j}\, \mathfrak{J}_\mathfrak{p}\zeta_i X_j$ is an R-subspace of V which spans V. Applying Lemma 9.8A with D = C = W, we see that there exist linearly independent vectors $Y_1, \cdots, Y_f$ in W such that $W = \mathfrak{J}_\mathfrak{p}Y_1 + \cdots + \mathfrak{J}_\mathfrak{p}Y_f$.

We denote the ideal (π) of $\mathfrak{J}_\mathfrak{p}$ by $\mathfrak{p}$ again. $R\pi = \pi R$ is a two-sided ideal of R, and the residue class ring $R/\pi R = \overline{R}$ is an algebra of rank n over the residue class field $\mathfrak{J}_\mathfrak{p}/\mathfrak{p} = \overline{\mathfrak{K}}$. We observe that $\overline{\mathfrak{K}} \cong \mathfrak{o}/\mathfrak{p}$, $\mathfrak{o}$ the ring of integers of $\mathfrak{K}$. We shall use $\overline{\alpha}$ to signify the residue class (mod πR) of the element α of R, and $\overline{c}$ to signify the residue class (mod $\mathfrak{p}$) of the element c of $\mathfrak{o}$.

The question arises as to the relationship between two R-subspaces W_1 and W_2 chosen as in Lemma 9.8C to span an A-space V. In the following argument, if W is an R-space with $\mathfrak{J}_\mathfrak{p}$-basis, we shall denote by $\overline{W}$ the $\overline{R}$-space $W/\pi W$. $\overline{W}$ determines a modular representation $\overline{Z}$ of $\overline{R}$ with coefficients in the field $\overline{\mathfrak{K}}$. We observe that, if $\alpha \rightarrow Z(\alpha)$ is the representation of R by $\mathfrak{J}_\mathfrak{p}$-matrices which is produced by the R-space W, then $\overline{\alpha} \rightarrow \overline{Z}(\overline{\alpha})$ is obtained by replacing the coefficients in $Z(\alpha)$ by their residue classes mod $\mathfrak{p}$.

Lemma 9.8D. _Let W_1 and W_2 be two R-subspaces of the A-space V which span V and have $\mathfrak{J}_\mathfrak{p}$-bases. Let $\overline{Z}_1$, $\overline{Z}_2$ be the modular representations of R which are given by the $\overline{R}$-spaces $\overline{W}_1$, $\overline{W}_2$. Then $\overline{Z}_1$, $\overline{Z}_2$ have the same modular irreducible constituents._

Proof. Since W_1, W_2 span V, then the representations Z_1, Z_2 of R are equivalent in $\mathfrak{K}_{\mathfrak{p}}$, that is, there exists a nonsingular matrix P, with coefficients in $\mathfrak{K}_{\mathfrak{p}}$, such that $Z_2 = P^{-1}Z_1P$, or

$$(8.2) \qquad Z_1(\alpha)P = PZ_2(\alpha).$$

We shall prove the theorem for any two related sets of $\mathfrak{J}_{\mathfrak{p}}$-matrices for which we have a relation of the form (8.2).

We may suppose that P is an $\mathfrak{J}_{\mathfrak{p}}$-matrix with not all coefficients divisible by π, for in the contrary case we could multiply (8.2) by a suitable power of π to obtain a matrix P of the desired form.

If det P is a unit of $\mathfrak{J}_{\mathfrak{p}}$, then, indicating residue classes mod $\mathfrak{p}$ by a bar, we have

$$\overline{Z}_1\ \overline{P} = \overline{P}\ \overline{Z}_2,$$

where $\overline{P}$ is nonsingular, and so $\overline{Z}_1$ and $\overline{Z}_2$ not only have the same irreducible constituents but are even equivalent.

If det P is not a unit, we may replace P by $P^* = MPN = \begin{Vmatrix} T & 0 \\ 0 & \pi Q \end{Vmatrix}$, where M, N are unimodular, T is a diagonal matrix of one or more rows, and det T is a unit, Q is an $\mathfrak{J}_{\mathfrak{p}}$-matrix. Z_1, Z_2 would be replaced by the representations $Z_1^* = MZ_1M^{-1}$, $Z_2^* = NZ_2N^{-1}$, and since $\overline{M}$, $\overline{N}$ are nonsingular, then, by the above remark, $\overline{Z}_1^*$, $\overline{Z}_2^*$ will have the same modular irreducible constituents as $\overline{Z}_1$, $\overline{Z}_2$, respectively. We split Z_1^*, Z_2^* according to the splitting of P^*,

$$Z_1^* = \begin{Vmatrix} A_1 & A_2 \\ A_3 & A_4 \end{Vmatrix}, \quad Z_2^* = \begin{Vmatrix} B_1 & B_2 \\ B_3 & B_4 \end{Vmatrix},$$

then

$$\begin{Vmatrix} A_1 & A_2 \\ A_3 & A_4 \end{Vmatrix} \begin{Vmatrix} T & 0 \\ 0 & \pi Q \end{Vmatrix} = \begin{Vmatrix} T & 0 \\ 0 & \pi Q \end{Vmatrix} \begin{Vmatrix} B_1 & B_2 \\ B_3 & B_4 \end{Vmatrix}.$$

Hence

$$A_1 T = T B_1 \qquad \pi A_2 Q = T B_2$$

$$A_3 T = \pi Q B_3 \qquad A_4 Q = Q B_4.$$

Then $\overline{Z}_1^* \simeq \begin{Vmatrix} \overline{A}_1 & \overline{A}_2 \\ 0 & \overline{A}_4 \end{Vmatrix}$, $\overline{Z}_2^* \simeq \begin{Vmatrix} \overline{B}_1 & 0 \\ \overline{B}_3 & \overline{B}_4 \end{Vmatrix}$. Here $\overline{B}_1 \simeq \overline{A}_1$ since det $T \not\equiv 0$ mod $\mathfrak{p}$. Further, the matrix sets A_4, B_4, which are of smaller degree than that of Z_1, Z_2, satisfy a relation of the form (8.2). We may then use an induction argument to prove the theorem.

Lemma 9.8E. Let $\bar{\mathfrak{l}} = \bar{R}\bar{\rho}$ be a nonnilpotent left ideal of the ring $\bar{R} = R/\pi R$, $\bar{\rho}$ an idempotent. Then there exists an idempotent e *of* R *such that $\bar{e} = \bar{\rho}$, and $\mathfrak{l} = Re$ is such that $\mathfrak{l}/\pi\mathfrak{l} \cong \bar{\mathfrak{l}}$.*

Proof. Let λ be an element of R such that $\bar{\lambda} = \bar{\rho}$. Then $\bar{\lambda}^2 = \bar{\rho}^2 = \bar{\rho}$, so that $\lambda^2 \equiv \lambda \pmod{\pi R}$, or $x = \lambda^2 - \lambda \,\varepsilon\, \pi R$. Now we take (cf. Theorem 2.4A) $\lambda_1 = \lambda + x - 2\lambda x$, then $x_1 = \lambda_1^2 - \lambda_1 = 4x^3 - 3x^2 \,\varepsilon\, \pi^2 R$. Continuing, with $\lambda_2 = \lambda_1 + x_1 - 2x_1\lambda_1$, we find $x_2 = \lambda_2^2 - \lambda_2 \,\varepsilon\, \pi^4 R$, and, similarly, $x_\nu = \lambda_\nu^2 - \lambda_\nu \,\varepsilon\, \pi^{2\nu} R$. Then $\lambda_{\nu+1} - \lambda_\nu = x_\nu(1 - 2\lambda_\nu) \,\varepsilon\, \pi^{2\nu} R$. It follows that the limit of the sequence

$$\lambda, \lambda_1, \lambda_2, \cdots, \lambda_\nu, \cdots$$

is idempotent. We denote this limit by e. Each $\bar{\lambda}_\nu$, and hence also $\bar{e}$, is $\bar{\rho}$. In the natural homomorphism of R on $\bar{R}$, $e \rightarrow \bar{e} = \bar{\rho}$, $Re \rightarrow \bar{R}\bar{e}$. An element $\alpha = \alpha e$ of Re maps on the zero element of $\bar{R}$ if and only if $\alpha \,\varepsilon\, \pi Re$. Then $Re/\pi Re \cong \bar{R}\bar{e}$, and our lemma is proved.

Let V be an irreducible A-space. By Lemma 9.8C, V is spanned by an R-space W with an $\mathfrak{I}_\mathfrak{p}$-basis. A contains $\mathfrak{K}_\mathfrak{p}$, R contains $\mathfrak{I}_\mathfrak{p}$, and V is a $\mathfrak{K}_\mathfrak{p}$-space, and W an $\mathfrak{I}_\mathfrak{p}$-space. Then $\bar{W} = W/\pi W$ is an $\bar{R}$ space, and so also a $\bar{\mathfrak{K}}$ space. Let, further, e be chosen as in Lemma 9.8D to correspond to $\bar{e}$ of the ideal $\bar{R}\bar{e}$ of $\bar{R}$. Then $Ae \supset Re$ is a nonnilpotent left ideal of A.

Lemma 9.8F. The rank with respect to $\mathfrak{K}_\mathfrak{p}$ of the set of A-homomorphisms of Ae into the A-space V equals the rank with respect to $\bar{\mathfrak{K}}$ of the set of $\bar{R}$-homomorphisms of $\bar{\mathfrak{l}} = \bar{R}\bar{e}$ into the $\bar{R}$-space $\bar{W} = W/\pi W$.

Proof. Let $\bar{\sigma}_1, \bar{\sigma}_2, \cdots, \bar{\sigma}_m$ denote a basis for the mappings of $\bar{\mathfrak{l}}$ on $\bar{W}$, the $\bar{\sigma}_i$ being linearly independent with regard to $\bar{\mathfrak{K}}$. Let $\bar{\sigma}_i(\bar{e}) = \bar{X}_i$, $X_i \,\varepsilon\, W$, then $\bar{e}\bar{X}_i = \bar{X}_i$. If $\bar{\sigma}$ is any mapping of $\bar{\mathfrak{l}}$ on $\bar{W}$, and $\bar{\sigma}(\bar{e}) = \bar{X}$, then it follows that $\bar{X}$ is linearly dependent on the $\bar{X}_i (i = 1, 2, \cdots, m)$. We next observe that $Y_i = eX_i \neq 0$, for otherwise $\bar{e}\bar{X}_i = 0$. Then $e \rightarrow Y_i = eY_i$, $\alpha e \rightarrow \alpha Y_i$ is an A-mapping of Ae on V, and this gives m linearly independent such mappings. We further observe that, since $X_i \,\varepsilon\, W$ and W is an R-space, $Y_i = eX_i \,\varepsilon\, W$. Let $\sigma(e) = Z$, $\sigma(\alpha e) = \alpha Z$ be any other mapping of Ae on V. We may suppose Z lies in W, for if it does not, we may replace Z by $\pi^\rho Z$, where π^ρ is a sufficiently high power of π that $\pi^\rho Z \,\varepsilon\, W$. Then $\bar{e} \rightarrow \bar{Z} = \bar{e}\bar{Z}$ gives a mapping of $\bar{\mathfrak{l}}$ on $\bar{W}$, and there exist elements $\bar{c}_1, \cdots, \bar{c}_m$, $c_i \,\varepsilon\, \mathfrak{o}$, such that

$$\overline{Z} = \overline{c}_1\overline{X}_1 + \overline{c}_2\overline{X}_2 + \cdots + \overline{c}_m\overline{X}_m.$$

It follows that

$$Z = c_1Y_1 + \cdots + c_mY_m + \pi Z_1, \; Z_1 \in W.$$

Here $eZ_1 = Z_1$, since $eZ = Z$, $eY_i = Y_i$. If $Z_1 \neq 0$, then we repeat the argument with the mapping $\sigma_1(e) = Z_1$, and obtain Z in the form

$$Z = (c_1 + \pi c_1')Y_1 + \cdots + (c_m + \pi c_m')Y + \pi^2 Z_2,$$

where $eZ_2 = Z_2$, $Z_2 \in W$. By continuing the process as far as necessary we obtain

$$Z = a_1Y_1 + a_2Y_2 + \cdots + a_mY_m,$$

with $a_i \in \mathfrak{J}_\mathfrak{p}$. Then $Y_1, \cdots, Y_m$ is a $\mathfrak{K}_\mathfrak{p}$-basis for the mappings of Ae into V, and our lemma is proved.

Let now $\overline{e}$, the unit element of $\overline{R}$, have the decomposition $\overline{e} = \overline{e}_1 + \overline{e}_2 + \cdots + \overline{e}_n$ into primitive mutually orthogonal idempotents. Let $B_1, \cdots, B_s$ denote the classes of isomorphic indecomposable left ideals $\overline{R}\overline{e}_i$, and let $\overline{R}\overline{e}_1, \cdots, \overline{R}\overline{e}_s$ be representative ideals of the classes $B_1, \cdots, B_s$. Then, if from the residue class $\overline{e}_i$ we choose e_i as in Lemma 9.8D, we obtain distinct ideals $Ae_1, \cdots, Ae_s$ of A. Let $V_1, V_2, \cdots, V_t$ be the distinct irreducible A-spaces. The V_i are the representation spaces of the essentially different irreducible representations of A in $\mathfrak{K}_\mathfrak{p}$. Let W_j be an R-space with $\mathfrak{J}_\mathfrak{p}$-basis which by Corollary 9.8C spans V_j, and $\overline{W}_j$ be the $\overline{R}$-space $W_j/\pi W_j$. Let $\overline{H}(.,.)$ be used to denote the rank of a set of homomorphisms; then from Lemma 9.8E we have

$$(8.3) \qquad H(Ae_j, V_i) = H(\overline{R}\overline{e}_j, \overline{W}_i),$$

where the rank on the left is with respect to $\mathfrak{K}_\mathfrak{p}$, and that on the right with respect to $\overline{\mathfrak{K}}$.

We assume in the following that A is semisimple. Let $\mathfrak{n}_1, \mathfrak{n}_2, \cdots, \mathfrak{n}_t$ denote the set of essentially different simple left ideals of A, and suppose the notation is so chosen that $\mathfrak{n}_i \cong V_i$ as an A-space. Then Ae_j is a direct sum of simple left ideals which are isomorphic to the ideals $\mathfrak{n}_i$, and we use the notation

$$(8.4) \qquad Ae_j \leftrightarrow \sum_{i=1}^{t} d_{ij}V_i \quad (j = 1, 2, \cdots, s)$$

to denote that d_{ij} of the simple left ideal summands in Ae_i are isomorphic to $\mathfrak{n}_i$. Let $\mathfrak{n}_i = Ag_i$, g_i idempotent. It follows that $H(Ae_j, V_i) = H(Ae_j, \mathfrak{n}_i) = q_i d_{ij}$, where q_i is the rank with respect to $\mathfrak{K}_\mathfrak{p}$ of the division algebra g_iAg_i.

Let $T_1, T_2, \cdots, T_s$ denote the distinct irreducible $\overline{R}$-spaces. Each T_i is isomorphic to one of the $\overline{R}$-spaces $\overline{R}\overline{e}_j/\overline{N}\overline{e}_j$ $(j = 1, 2, \cdots, s)$. We write

$$(8.5) \qquad \overline{W}_i \leftrightarrow \sum_{j=1}^{s} \tilde{d}_{ij} T_j$$

to denote that $\tilde{d}_{ij}$ of the factor groups in a composition series for $\overline{W}_i$ are isomorphic to T_j. Since $\overline{\mathfrak{R}}$ is a Galois field, and hence perfect, we may assume that the completely primary ring $\overline{e}_j\overline{R}\overline{e}_j$ over $\overline{\mathfrak{R}}$ can be written as a direct sum of its radical and a division algebra $\overline{h}_j$. Let r_j denote the rank of $\overline{h}_j$ over $\overline{\mathfrak{R}}$. Then, by Theorem 9.5A, $H(\overline{R}\overline{e}_j, \overline{W}_i) = \tilde{d}_{ij}r_j$. We now have

$$(8.6) \qquad q_i d_{ij} = \tilde{d}_{ij}r_j.$$

We next observe that Re_j is an R-space with $\mathfrak{J}_\mathfrak{p}$-basis which spans $V = Ae_j$. For, if $\zeta_1, \cdots, \zeta_n$ be an $\mathfrak{J}_\mathfrak{p}$-basis for R (see Corollary 9.8B), then $Re_j = \mathfrak{J}_\mathfrak{p}\zeta_1 e_j + \cdots + \mathfrak{J}_\mathfrak{p}\zeta_n e_j$ and, by Lemma 9.8A, it follows that Re_j has an $\mathfrak{J}_\mathfrak{p}$-basis. On the other hand, $W = \Sigma_{i=1}^{t} d_{ij}W_i$ is also an R-space which spans V and has an $\mathfrak{J}_\mathfrak{p}$-basis. Then, by Lemma 9.8D, $\overline{R}\overline{e}_j$ and $\overline{W} = \Sigma_{i=1}^{t} d_{ij}\overline{W}_i$ have the same modular composition spaces, so that

$$(8.7) \qquad \overline{R}\overline{e}_j \leftrightarrow \sum_{i=1}^{t}\sum_{k=1}^{s} d_{ij}\tilde{d}_{ik}T_k.$$

This shows that c_{kj}, the number of composition spaces of $\overline{R}\overline{e}_j$ which are isomorphic to T_k is

$$c_{kj} = \sum_{i=1}^{t} d_{ij}\tilde{d}_{ik},$$

or, from (8.6),

$$(8.8) \qquad c_{kj} = \sum_{i=1}^{t} \frac{d_{ij}d_{ik}q_i}{r_k}.$$

The formula (8.8) is the relation we have sought between the Cartan invariants of $\overline{R} = R/\pi R$ and the decomposition numbers d_{ij}.

INDEX

CPSIA information can be obtained
at www.ICGtesting.com
Printed in the USA
BVHW031509050321
601766BV00010B/47